AGRICULTURE ISSUES AND POLICIES

PESTICIDES

CHARACTERISTICS, USES AND HEALTH IMPLICATIONS

Agriculture Issues and Policies

Additional books in this series can be found on Nova's website under the Series tab.

Additional e-books in this series can be found on Nova's website under the e-book tab.

Chemical Engineering Methods and Technology

Additional books in this series can be found on Nova's website under the Series tab.

Additional e-books in this series can be found on Nova's website under the e-book tab.

AGRICULTURE ISSUES AND POLICIES

PESTICIDES

CHARACTERISTICS, USES AND HEALTH IMPLICATIONS

AARÓN J. ABREGO GOMEZ
AND
EDUARDO M. LUGO DE ORTEGA
EDITORS

Nova Science Publishers, Inc.
New York

For permission to use material from this book please contact us:
Telephone 631-231-7269; Fax 631-231-8175
Web Site: http://www.novapublishers.com

NOTICE TO THE READER

Library of Congress Cataloging-in-Publication Data

Pesticides : characteristics, uses and health implications / editors: Aarsn J. Abrego Gomez and Eduardo M. Lugo de Ortega.
p. cm.
Includes index.
ISBN 978-1-62081-577-9 (hardcover)
1. Pesticides--Environmental aspects. 2. Pesticides--Health aspects. I. Abrego Gomez, Aarsn J. II. Lugo de Ortega, Eduardo M.
QH545.P4P4798 2012
363.738'498--dc23

2012013583

Published by Nova Science Publishers, Inc. † New York

Contents

PREFACE

In this book, the authors present current research in the study of the characteristics, uses and health implications of pesticides. Topics discussed include the evaluation of pesticide-induced DNA damage and oxidative stress on human and wildlife populations in Argentina; analytical methods of various families of pesticides in biological matrices; implications of pesticide use in the olive sector; pesticides and endangered sea turtles; monitoring emissions and atmospheric degradations of pesticides in the atmosphere; and adverse effects of pesticides in human health.

Chapter 1 - For a long time toxicity of pesticides was directly associated to acute effects, while nowadays the main problem is related to massive application and the effects induced by chronic exposure to pesticides and complex mixtures, even at very low concentrations.

Among the wide range of damages cause by pesticides on health, genotoxicity and oxidative stress (OE) are considered biologically relevant and highly informative as early warnings of the impacts on natural populations. OE is the result of an imbalance between free radicals formation and antioxidant defense mechanisms. In this condition, the excess in reactive molecules can lead to direct or indirect damage to proteins, lipids and DNA. The final fate of these alterations is uncertain, but they could affect the normal function of physiological processes, with consequences at cellular, individual and population level.

In the present chapter, the authors present an evaluation of the environmental situation of an area highly exposed to pesticides (Santa Fe province, Argentina), including studies made in wildlife species as well as in human populations directly and indirectly exposed to pesticides and mixtures. They use biomarkers of genotoxicity (Comet assay and Micronucleus test), and oxidative stress (damage to lipids: TBARS, to DNA: Comet assay modified with bacterial enzyme FPG, and antioxidant defense: CAT, SOD y GSH). In human populations, confounding factors were also analyzed.

The results obtained at different level could be highly valuable to estimate the real impact of chronic exposure of human and wildlife populations to pesticides mixtures in this region of Argentina.

Chapter 2 - Pesticides are among substances that have been identified as potential Endocrine Disrupting Chemicals (EDCs). The availability of validated exposure data is a critical component for assessing the causal relationships between exposure to EDCs and health effects. Although more and more data are available on these compounds, lots of questions about their ways of action are not answered yet and it is important to keep on studying them. The development of reliable multi-residue analytical methods is necessary to undertake this kind of study, especially on biological matrices such as plasma, serum, whole blood, urine, tissues, etc.

One of the main challenges in the development of analytical methods in biological matrices is the small amount of sample available. Therefore, special attention must be paid to the miniaturization of each step of the analytical method. Besides, as the number of samples in biomonitoring studies is very high, analytical methods developed have to be as fast as possible. Finally, studies on the behavior of pesticides in biological matrices not only require great sensitivity on a limited number of compounds but also information on untargeted molecules, such as their metabolites.

This chapter aims at giving an overview of current analytical methods of various families of pesticides in biological matrices, focusing on miniaturized sample preparation. The development of untargeted analytical methods and current data on pesticides transformation products will also be discussed.

Chapter 3 - The application of phytosanitary products is one of the main employed procedures to solve the problems of plagues and diseases in olive groves. As a result, there is an increasing unease related to the possible presence of residues from these products in olive-derived foods. Recent studies have shown that these residues can be found in olives and olive oil (mainly insecticides, fungicides and herbicides) due to the lipophillic nature of the used plaguicides, usually at levels below the maximum residue limits (MRLs) established. Due to the fact that virgin olive oil production has increased in recent years and that it is being exported to different countries, exhaustive quality controls are required. In this sense, the quantification of pesticides residues is one of the most demanded analyses nowadays.

Olive farmers employ a wide range of pesticides in the plantations, sometimes without full understanding of their impact on human health and the environment. This constitutes a risk of these pesticides being incorporated in both: (1) rivers and reservoirs due to the soil that is eroded in heavy rains, polluting surface waters, etc.; and (2) olives and, consequently, olive oil and table olives. Different regulations regarding MRLs in table olives and olive oil have been established by the European Union and the Codex Alimentarius of the Food and Agriculture Organization of the United Nations. There are alternative methods available for the analysis of pesticides in different kinds of samples. However, the most common methods of analysis are Gas Chromatography (GC) and Liquid Chromatography (LC).

Our research group has performed thorough investigation over the analysis of pesticides in soils, olive fruits and virgin olive oils. In particular, the authors have developed different GC analytical methods with mass spectrometric detection for the analysis of all the main pesticides used in olive groves. In this way, sensitive and reliable procedures can be implemented for the quality control of table olives and olive oils in the industry. After describing the process for olive oil production, together with the common plagues and pesticides used in the olive groves, the methodologies developed by the authors' research group will be described. Finally, the results obtained will be clearly explained.

Chapter 4 - Sea turtles are considered of increasing interest as potential bioindicators for pollution in marine ecosystems. In addition, all species of sea turtles are included on the Red List of the World Conservation Union, and all the threats for their survival must be investigated, including the health implications of pesticides.

This chapter is a review of the international research made on pesticides and sea turtles. The majority of the studies reported baseline levels of organochlorine (OC) pesticides in several species of sea turtles from different geographical areas. Several sea turtle samples were used to determine the concentration of OC pesticides: eggs and chorioallantoic membranes, tissues, and blood. Contamination of sea turtle eggs by pesticides, effects on embryonic development, maternal transfer, and even implications for public health (consumption of sea turtle eggs) were investigated. Several specific studies investigated the effects of pesticides on sexual differentiation, sexual steroid-binding proteins, and immunity. Other relevant studies tried to establish an association between OC pesticides and clinical health parameters and/or diseases.

Chapter 5 - The agricultural production, in terms of quality, productivity, and management, improved with the extensive application of pest control technologies, being today's agricultural practices heavily dependent on the use of them. The *pesticides* are the most important tool in intensive crop-based agriculture and in some public health applications. They are chemical or biological products used to prevent diseases, kill, repel, control plagues or interrupt the growth of plants, including herbicides, fungicides, acaricides, avicides, and insecticides. Different families of chemical compounds and formulations have been proposed to increase their efficacy using a wide range of action modes. The most important families are organophosphates, carbamates, organochlorines, chloroacetanilines, anilines, chloroanilines, dinitroanilines, substituted ureas, and triazines, presented as solids, solutions, suspensions or emulsions. The pesticides can affect energy metabolism or growth in plants and nervous system, muscle action or growth regulation in animals. Unfortunately, the fraction of the dosage of applied pesticide enters into the different *environmental compartments* is quite important.

The distribution of a pesticide in the environment is controlled by their physical-chemical properties. For instance, Henry Law constant (the distribution coefficient between water and air) defines the predominant movement via air or via water phase (Bromilow 2005). The exposition risks come from pesticides present in the soil, waters, foods, and also the atmosphere. Therefore, their *adverse effects* on human health and the environment have changed the initial indiscriminate use of synthetic organic pesticides. Health effects in humans and other organisms are caused by skin contact, inhalation, and ingestion.

The *atmosphere* represents the largest environmental compartment, one of the first destinations after the pesticide releases, and a relevant exposition source for living organisms. The human exposure to pesticides from the atmosphere is classified in occupational (farmers and related workers) and general population as the result of their high persistence or the exposition of degradation pesticide products. The accurate risk assessment is complicated because the adverse effects depend on factors as periods and levels of exposure. The characteristics of the atmosphere allow a fast movement, for instance the diffusion in air approximately 10^4 times faster than in water. There are several mechanisms for exchanging with the surface components (soils, water, plants or animals) through the boundary layer – lower area of troposphere in contact with terrestrial surface.

The emission is the introduction of a pesticide into the atmosphere. When atmospheric contamination phenomenon occurs, it is a consequence of either a specific, identifiable source or place (point-source pollution) or it can be a combination of different sources in a wide area (non-point-source pollution). The emission of pesticides is produced during their manipulation, application or post-application (Table 1).

The reduction of the emission during the *manipulation* can be achieved with a correct use. The labeling of pesticide products should include warning statements that the pesticide handler should heed to avoid negative effects. Farmers or pest-control workers must take precautions to make sure that the pesticides remain in a sealed container until they are released into the application site, which also must be controlled to prevent the vapor from escaping or accidental spilling. Personal protective and application equipment requires a complete washing to reduce toxicological effects in workers, but these operations also produce the losses of pesticides to the indoor or outdoor atmosphere.

During the *application*, an important fraction of the pesticide always is lost and directly passes to the air. For instance, spraying by aircrafts, vehicle-mounted applicators, or manual equipments generate droplets with different sizes and some of them do not reach the target land surface or leaves. Losses during the application of pesticides can range to 50%, or even more in some cases. The main factors that affect pesticide air emission are the manner in which the pesticide is applied; concretely, the spray characteristics (e.g., size of nozzles), the equipment used (e.g., manual or vehicle), the formulation used (e.g., viscosity, volatility), the environmental conditions (e.g., wind speed, temperature, humidity, atmospheric stability), and finally the operators skills. Pesticides move away from the release site through wind (outdoor applications) or through air currents generated by ventilation systems (indoor applications).

The *post-application* emission has an important contribution. Even surface dissipation processes like wash-off due to rain or run-off are done, the atmospheric emission of pesticide are continuously produced until complete animal/plant uptake or soil adsorption (target or not-target). Their volatilization from the treated surface represents the most important input into the troposphere for days or even weeks after the application. Depending on the pesticide used and the climatic conditions, the volatilization can reach high percentage of the application dose. Vapor pressure, water solubility, and Henry's law constant could be considered the best parameters to describe the volatilization of a pesticide from applied surface or water.

Chapter 6 - Pesticides are among the most widely used chemicals in the world, and are also one of the most dangerous to human health. According to the Stockholm Convention on Persistent Organic Pollutants, 13 of the 21 most dangerous and persistent organic chemicals are pesticides. The Food and Agriculture Organization (FAO) has defined pesticides as: "*any substance or mixture of substances intended for preventing, destroying or controlling any pest, including vectors of human or animal disease, unwanted species of plants or animals causing harm during or otherwise interfering with the production, processing, storage, transport or marketing of food, agricultural commodities, wood and wood products or animal feedstuffs, or substances which may be administered to animals for the control of insects, arachnids or other pests in or on their bodies. The term includes substances intended for use as a plant growth regulator, defoliant, desiccant or agent for thinning fruit or preventing the premature fall of fruit. Also used as substances applied to crops either before or after harvest to protect the commodity from deterioration during storage and transport.*"

Pesticides are designed to kill, suppress or alter the life cycle of pests, such as insects, rodents, fungi and unwanted plants (weeds). In the public health sector, they are used to control vectors of diseases while in agriculture they are used to kill pests. However, they are not selective, and can also kill or cause harm to non-target organisms. The most common types of pesticides include: herbicides, insecticides, fungicides, rodenticides, pediculicides and biocides.

Chapter 7 - The pesticides are one of the chemical compounds most widely used in the world. The pesticide description commonly referred substances that control organisms which destroy plant life and interfere with the food chain, and which act as vectors for disease organisms to man and animals. This generic definition is frequently extended, rather unsatisfactorily and inaccurately, to cover other chemicals used on plants, such as growth regulators. Pesticides can be classified in different ways, such as by their target, chemical nature, physical state, and mode of action. Classification based on the target (insecticides, herbicides, fungicides, and rodenticides) and chemical structure (organophosphates, carbamates, organochlorines, organochlorines, pyrethroids, phenols, morpholines, chloroalkylthiols, organometallics, azoles, bipyridilium, ureas, anilines, chloronitrile) are the most widely known. Due to their widespread use, a large number of poisonings occur every year in occupationally-exposed human populations, and in the general population appear to regard exposure to pesticides from residues in food and, perhaps, water of greatest concern, there are multiple other sources of exposure which can compound with those from residues. These include hand-to-mouth contact from pesticides used within buildings, veterinary medicines used against domestic pets (flea sprays), and contamination of food and working surfaces from the residential use of pesticides (control of insects). Thus, although the oral route is probably the major route of exposure for the general public, the skin and eyes probably are also significant, and inhalation the least through the consumption of contaminated food and drinking water. The types of adverse health effects that have been documented in exposed workers with pesticides as a group have included acute toxic effects (mirroring the mechanism of toxic action), primary irritancy (the skin and eyes), sensitization (mainly allergic contact dermatitis, but on occasion respiratory sensitization as been described), peripheral and central neurotoxic effects, myonecrosis and in epidemiology studies, there have been suggestions of cardiovascular toxicity, reproductive and developmental toxicity, endocrine alterations, genotoxicity, oncogenicity and immunotoxic effects among other. This review article considers the chemistry, characteristics, uses and health effects of representative groups of pesticides.

Chapter 8 - The use of pesticides for pest control in non-agricultural or urban amenity areas has often been discussed to potentially lead to different fate and environmental as well as human side-effects than when used in agriculture. However, relatively little attention has been directed so far to the risk assessment of pesticides in urban as compared to agricultural areas, both in terms of research efforts and legislation. The present study was initiated as a first step to increase the authors' knowledge related with pesticide use in urban areas. To this end, information on main pests, pesticides used to counteract these, and application techniques were gathered through surveys of pesticide operators in a Municipality in Portugal. A multicompartment model, leaching index and potential for particle-bound transport were applied by using the physicochemical properties of these pesticides to evaluate the environmental fate of the pesticides used in the present case study and obtain an indication of the compartments potentially at risk. These analyses will aid in selecting pesticides to be monitored in different compartments (drainage water, surface water, groundwater, soil) in future studies aiming at increasing our understanding of pesticide fate and their possible side-effects (applicatoroperator, bypassers, urban dwellers, environment) in urban settings. In addition, several constraints in knowledge of the operators related with pesticide application and storage were identified that may result in risk situations for both the environment and human health.

In: Pesticides: Characteristics, Uses and Health Implications
Editors: A. Abrego Gomez and E. Lugo de Ortega
ISBN: 978-1-62081-577-9

Chapter 1

EVALUATION OF PESTICIDE-INDUCED DNA DAMAGE AND OXIDATIVE STRESS ON HUMAN AND WILDLIFE POPULATIONS IN SANTA FE PROVINCE (ARGENTINA)

G. L. Poletta[1,2,3,4,*], M. F. Simoniello[2], M. Porcel de Peralta[2], E. Kleinsorge[2], P. Siroski[1,4] and M. D. Mudry[3,4]

[1]Proyecto Yacaré - Laboratorio de Zoología Aplicada:
Anexo Vertebrados (FHUC – UNL / MASPyMA), A. del Valle,
Santa Fe, Argentina
[2]Cat. Toxicología, Farmacología y Bioquímica Legal,
Facultad Bioquímica y Ciencias Biológicas, UNL, Santa Fe, Argentina
[3]Grupo Investigación Biología Evolutiva, IEGEBA (CONICET-UBA),
Buenos Aires, Argentina
[4]CONICET, Argentina

ABSTRACT

For a long time toxicity of pesticides was directly associated to acute effects, while nowadays the main problem is related to massive application and the effects induced by chronic exposure to pesticides and complex mixtures, even at very low concentrations.

Among the wide range of damages cause by pesticides on health, genotoxicity and oxidative stress (OE) are considered biologically relevant and highly informative as early warnings of the impacts on natural populations. OE is the result of an imbalance between free radicals formation and antioxidant defense mechanisms. In this condition, the excess in reactive molecules can lead to direct or indirect damage to proteins, lipids and DNA. The final fate of these alterations is uncertain, but they could affect the normal function of physiological processes, with consequences at cellular, individual and population level.

---

* Corresponding author. Tel.: +54 342 4575215 int.155; Fax: +54 342 4575221; E-mail address: gpoletta@fbcb.unl.edu.ar.

In the present chapter, we present an evaluation of the environmental situation of an area highly exposed to pesticides (Santa Fe province, Argentina), including studies made in wildlife species as well as in human populations directly and indirectly exposed to pesticides and mixtures. We use biomarkers of genotoxicity (Comet assay and Micronucleus test), and oxidative stress (damage to lipids: TBARS, to DNA: Comet assay modified with bacterial enzyme FPG, and antioxidant defense: CAT, SOD y GSH). In human populations, confounding factors were also analyzed.

The results obtained at different level could be highly valuable to estimate the real impact of chronic exposure of human and wildlife populations to pesticides mixtures in this region of Argentina.

1. History and Transformation of Agriculture in Argentina

Agriculture in Argentina began approximately 150 years ago in the Pampean region. This activity has been a prime contributor to economic and social development of the country (Lantieri et al., 2009). Worldwide, the so-called "Green Revolution", with the incorporation of new machinery, the massive use of agrochemical products and fertilizers, and the use of improved seeds, helped to double and even triple the yields of the most important grains (Lantieri et al., 2009). In Argentina, soybeans occupied 36,000 hectares in 1970, compared with 1.7 million hectares in Brazil and over 17 million in the United States. By 1990, 5.0 million hectares were already planted with soybeans in Argentina, with a growing rate of nearly 8 percent per year (Lantieri et al., 2009). Since then, there was an outstanding increase in farming activity known as "agriculturalization", characterized by a strong and continuous increase in the land area dedicated to crop growing and change in land use, including the conversion of large areas of native ecosystems into farmlands, with the highest rates of deforestation in South America (0.8%/year) (Lantieri et al., 2009; Navarrette et al., 2009). Agricultural crops also steadily replaced stockbreeding and dairy-farming (Aizen et al., 2009). Then, the process started to expand towards extra-Pampean regions with the transfer of capital, technology, and new cultivation schemes to non-agricultural areas (Stratta Fernández & Ríos Carmenado, 2010).

The introduction of genetically modified (GM) soybean resistant to glyphosate was the most important technological innovation in Argentina´s agriculture since 1996. An extensive agricultural model developed, mainly in response to international prices (Lanteri, 2009), based on glyphosate-resistant transgenic soybean farming, no-tillage (NT) system and the intensive use of fertilizers and pesticides. By 2011, there was more than 20 million hectares cultivated with transgenic soybean in the country (SAGPyA, 2011), continuously expanding towards natural ecosystems (Aizen et al., 2009).

NT systems allowed reducing damage caused to the soil by erosion, and improved water storage. However, this caused an increase in soybean monoculture and in the use of agrochemicals (fertilizers, herbicides, insecticides and fungicides) applied for different purposes (Hang et al., 2011). The pair GM soybean / NT system became the most widespread agricultural system but the sustainability of this association has been repeatedly questioned, particularly in the areas of extra-Pampa (Navarrette et al., 2009).

Nowadays, soybean occupies a wide ecological region in Argentina, from 23° to 39° Lat. S, with the higher concentration in the Pampean region (Figure 1).

2. Santa Fe Province

Santa Fe province is situated in the center-east region of Argentina (from 28° 00' 25" to 34° 25' 56" Lat. S and 62° 06' 44" to 58° 54' 26" Long. W) (Figure 2). Historically, this province has been one of the main agriculture centers of Argentina, representing more than 21 % of the cultivated area of the country. The main crops are soybean, sunflower, maize, wheat, and rice. Nowadays, Santa Fe is one of the main national producers of soybean (Figure 1), contributing with 30 % of the production (Penna & Lema, 2002; Lanteri, 2009). The area cultivated with this crop in the province rose from 1,931,700 hectares in 1992 to 3,107,737 in 2010 (SIIA, 2011).

As in other provinces of Argentina, the expansion of cultivated lands replaced stockbreeding and dairy farming and also invaded natural ecosystem areas, with a high deforestation rate. More than 50% of transgenic soybean cultivated in the last decade has extended over areas which corresponded previously to native forest (Paruelo et al., 2006).

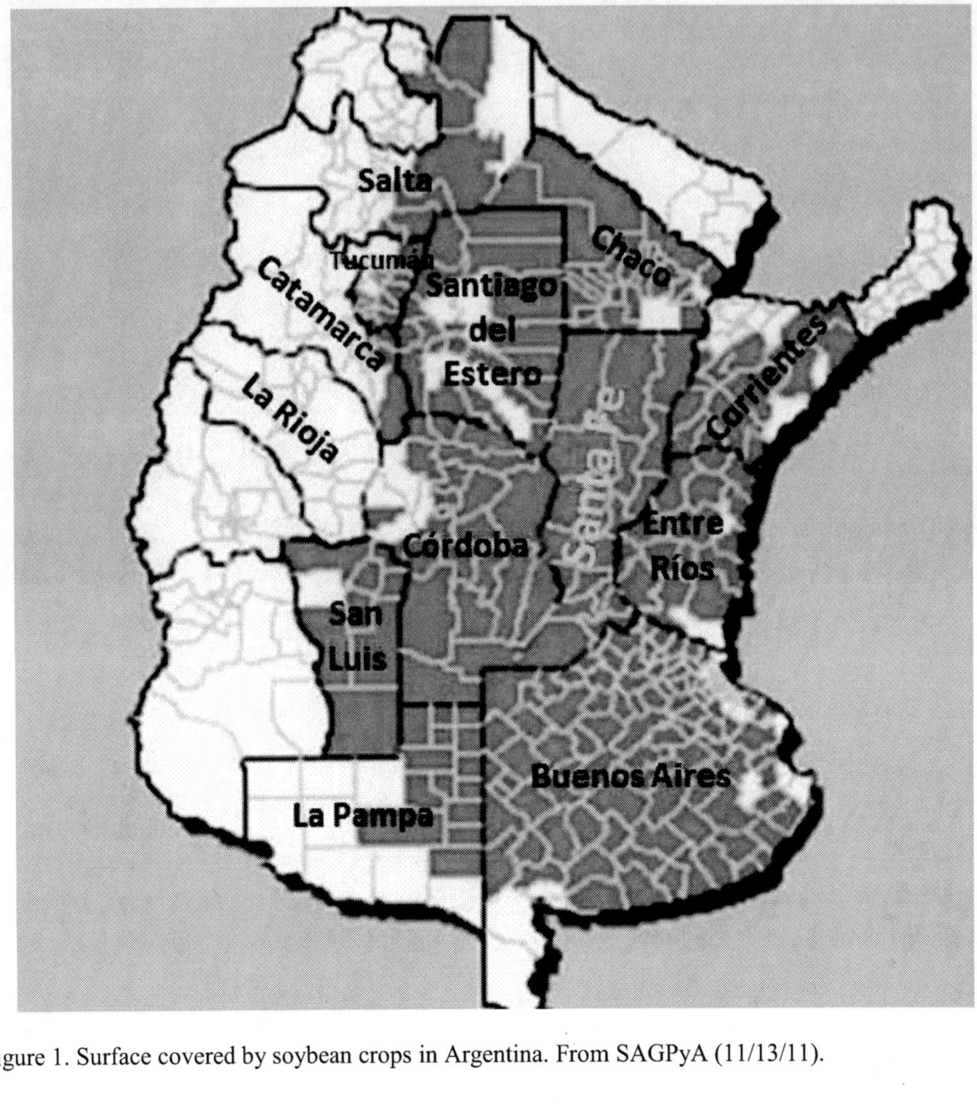

Figure 1. Surface covered by soybean crops in Argentina. From SAGPyA (11/13/11).

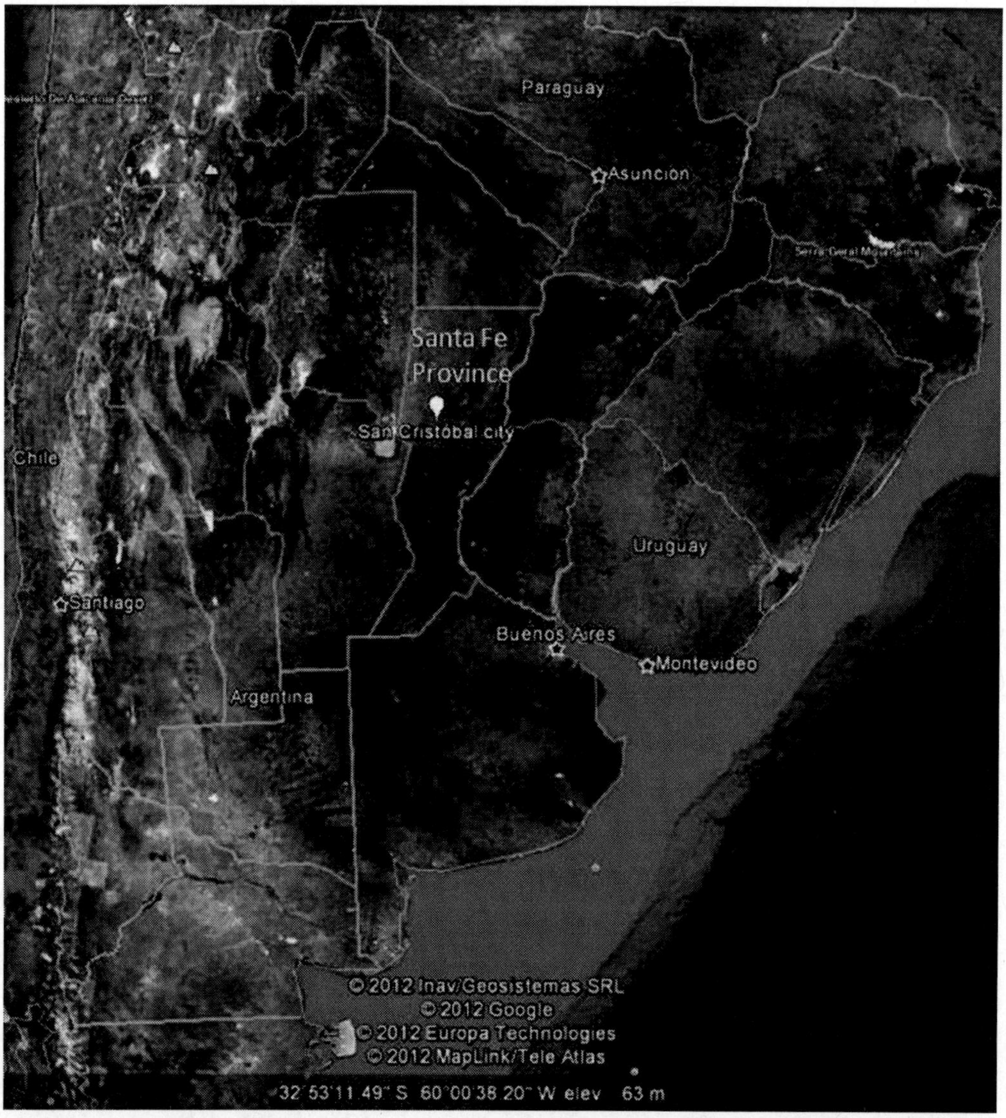

Figure 2. Santa Fe province is located in the center-east region of Argentina. It can be seen the study site, San Cristóbal department, situated in the center-north region of Santa Fe province.

3. Pesticide Use and Health Implications

The main environmental problem associated with agriculture is diffuse pollution. The fate of pesticides, once they have been applied, is highly complex and dependant on many interacting factors, such as the properties of the compound, soil properties, environmental conditions and the site characteristics. Water and sediment are transported by erosion, and both in liquid and solid phase, pesticides can be transported too. Transport by leaching, migration, runoff, sediment and drift are the main causes of pollution associated with agricultural activity (Peruzzo et al., 2008; Hang et al., 2011).

The use of glyphosate (phosphonoglycine) in Argentina increased exponentially since the introduction of genetically modified soybean resistant to glyphosate in the years 1996-1997. Its fast growth was due to the wide spectrum of weeds controlled, high efficiency and low cost. Another aspect that contributed to its increased use was the expansion of the area planted towards semi-arid regions of Argentina on NT systems (Navarrete, 2009). Moreover, the use of glyphosate is not limited to soybeans, as it is applied as a total herbicide in a wide number of crops. In the last season in Argentina, of the total 200 million liters of pesticides applied in agriculture, more than 170 correspond only to glyphosate-based formulations, followed by the insecticides cypermethrin and endosulfan (CASAFE, 2011). Recent studies made in the pampasic region of Argentina revealed glyphosate residues in surface water after spraying, as a consequence of drift and runoff (Peruzzo et al., 2008).

Pesticides may cause acute and chronic effects on non-target organisms before they are broken down into harmless compounds, thus, their persistence in soil and water is a key determinant of their environmental impact (Hang et al, 2011). They may produce deep consequences in ecosystems, including death and lesions on organisms, suppression of immune system, disruption of endocrine system, reproductive inhibition or failure, teratogenic and carcinogenic effects, cellular and molecular alterations including DNA damage, among others (Donald, 2004; McLaughlin & Mineau, 1995; McCauley et al., 2006).

Several studies have shown that sub-lethal and sub-organismal level effects, such as DNA damage, small turnover rate of proteins, etc., influence energy metabolism, fitness and reproductive success, leading to population level effects (Bickham, et al., 2000; Tsui & Chu, 2003). The increase in mutation rate would lead to an increase in genetic load and a reduction in adaptive potential, with the consequential elimination of susceptible genotypes (Grisolia, 2005; Udroiu, 2006). Therefore, increase genomic instability has been suggested to play important roles in decreased fitness of the populations (Saura-Mas et al., 2002, Jha et al., 2008).

Many pesticides can produce toxicity to organisms by the production of reactive oxygen species (ROS) (Azqueta et al., 2009; Franco et al., 2009). ROS include free radicals and other highly reactive forms of oxygen (e.g. hydrogen peroxide, superoxide anion radical, singlet oxygen, hydroxyl radical, nitric oxide and peroxynitrite) (Azqueta et al., 2009). Cell damage caused by an excess of oxidants, including ROS, has been defined as oxidative stress. In excess, ROS can overwhelm the normal antioxidant buffering capacity of the cell, leading to significant damage to cellular components, including proteins, lipids and DNA. Thus, ROS contribute to formation of mutations and, subsequently, to the etiology of degenerative diseases such as cancer (Üner et al. 2006; Azqueta et al., 2009; Limón-Pacheco & Gonsebatt, 2009). Organisms protect themselves from such damage with both enzymatic and non-enzymatic antioxidant defenses. Three levels of protection have been considered: 1) prevention of ROS formation, 2) termination of the ROS using free radical scavengers or antioxidant enzymes, and 3) repair of damaged cellular components (Storey, 1996).

Lipid peroxidation may be the first step of cellular membrane damage and can be induced by environmental pollutants as pesticides (Stark, 2005). This occurs in an initiation stage through the extraction of hydrogen from membrane lipids by ROS such as the hydroxyl radical (HO) (Goetz & Luch, 2008). Alteration in antioxidant enzymes is also a consequence of oxidative stress.

4. Biomarkers and Biomonitoring

By this decade, biomarker studies, used in the evaluation of environmental health as n indicators of toxic effects of environmental pollutants, have become essential. Changes in biochemical level are the "early warning" responses of an organism to environmental alterations and are critically important. One of the most essential features of molecular or cellular biomarkers is that they give a potential forecast about changes on upper levels of biological organizations, at the population, community, and ecosystem levels. As a consequence, biomarkers can be taken as short-term indicators of biological effects that will be seen in the long term (Çok et al., 2011). Some biomarkers can be used to evaluate the effects that arise from exposure to genotoxic xenobiotics that affect the DNA. In recent years, much interest has been shown in measuring DNA single-strand breaks as a sensitive indicator of genotoxicity. The quantity of single-strand breaks is associated with different types of environmental pollutants that have a mutagenic and carcinogenic character (Mitchelmore & Chipman, 1998). Single-cell gel electrophoresis (SCGE) or Comet Assay is one of the most widely applied biomarkers in the investigation of DNA damage, due to its simplicity, quickness of results, and high sensibility to detect low levels of damage in individual cells of various types of organisms, from invertebrates to vertebrates, including fish, amphibians, reptiles and mammals (Jha, 2008; Poletta et al., 2008; Dawhan et al., 2009; Frenzilli et al., 2009;). It has found a place in biomonitoring studies as a useful tool in the evaluation of various types of DNA damage and its repair (De Andrade et al., 2004; Piperakis et al., 2006; Cornetta et al., 2008).

The main outcome of the Comet assay is the detection of excessive DNA strand breaks, a type of lesion of great importance that reflects the recent status of pollution, has a temporal nature, and may be removed by the DNA repair system (Collins, 2004). But chronic exposure to pollutants can lead to accumulation of DNA strand breaks, as the DNA repair capacity of cells, an important mechanism to protect DNA integrity, is low in some species. If not repaired, these DNA lesions can initiate a cascade of biological consequences at the cellular, organ, organism, and, ultimately, at the population levels (Shugart, 2000; Jha, 2008).

The Micronucleus (MN) test, has a highly predictive value to detect clastogenic and aneugenic effects of compounds in human and wildlife populations environmentally exposed. Micronuclei represent loss in chromatin as a result of damage to either chromosome structure (fragmentation) or the mitotic apparatus (Schmid, 1975; Fenech, 2000).

On the other hand, the production of reactive oxygen species (ROS) and enhancement in the activities of antioxidant enzymes such as catalase (CAT) have been proposed as indicators of pollutant-mediated oxidative stress (Rossi et al., 2011). ROS, such as hydrogen peroxide (H_2O_2), hydroxyl radicals ($OH^\cdot$) and superoxide anions (O_2^-) have been shown to damage DNA and other cellular components, resulting in DNA degradation, protein denaturation, and lipid peroxidation. DNA damage induced by oxygen radicals occurs by oxidative modification of the bases in nucleic acids and by scission of the DNA strands (Rossi et al., 2011). Measurements of lipid peroxidation products, such as malondialdehyde (MDA), as well as modifications in endogenous oxygen free radical (OFR) scavengers, including superoxide dismutase (SOD) and catalase (CAT), are used as effective biomarkers to study OFR effects. With regard to the detection of oxidized bases, the Comet assay modified by using repair endonuclease, such as Formamidopyrimidine - DNA glycosylase (FPG), has demonstrated to be the most sensitive and accurate technique for measurement of oxidized bases (Collins, 2009).

PESTICIDES AND HUMAN HEALTH

The wide use of pesticides in agricultural and other settings results in continuing human exposure, showed by epidemiological studies associated with risks to health (Alavanja et al., 2004). Chronic exposure is most often a problem in occupational settings and can increase the risk of developmental and reproductive disorders, immune system diseases, endocrine disruption, impaired nervous-system function, and development of certain cancers (WHO, 1990; Yañez et al., 2002; Bassil et al., 2007).

Chronic exposure to pesticides involves the existence of complex mixtures of different types of chemicals: active ingredients and by-products, such as impurities, solvents and other compounds produced during the storage procedure, present in technical formulations. Moreover, although inert ingredients have no pesticide activity, they may be biologically active and sometimes the most toxic component of a pesticide formulation. Each active ingredient has a specific mode of action for controlling a pest, and has its own possible side effects on humans exposed to it. The exposure to different pesticides or pesticide mixtures, either simultaneously or in series, makes it difficult to identify the effects of each one separately. Biomonitoring could be used as a tool of vigilance to help understanding clinical problems or to evaluate a certain occupational or environmental exposure. However, most epidemiological studies assessing health and environmental risks have focused on the assessment of single exposures without addressing the effects of mixtures (Lantieri et al., 2009). The synergistic effect of mixtures on the etiology of disease is of increasing concern for public health.

Biomonitoring of Human Populations Exposed to Pesticide Mixtures in Santa Fe Province (Argentina)

Introduction

In third world countries, pesticide poisoning is a major problem primarily due to precarious and unsafe pesticide application and handling practices. In countries where economies are greatly dependent on agriculture, such as Argentina, the good management, use, and disposal of pesticides become a health and environmental issue in agricultural settings (Lantieri et al., 2009).

Genotoxicological biomonitoring of human populations is a useful tool to estimate the genetic risk posed by an integrated exposure to complex mixtures of chemicals. In different countries there are numerous studies of genotoxicological biomonitoring in populations exposed to pesticides (Carbonell et al., 1990; Dulout et al., 1992; Bolognesi et al., 1993; Scarpato et al., 1996; Amr, 1999; Falck et al., 1999; Gomez Arroyo et al., 2000; Lander et al., 2000; Piperakis et al., 2006; Remor et al., 2009). However, more studies are still needed in Argentina (Simoniello et al., 2008, 2010; Mañas et al., 2009).

Glutathione S-transferases (GSTs) belong to a superfamily of detoxification enzymes that provide critical defenses against toxicants (Board et al., 1990; Mannervick et al., 1992). Recent studies indicate that the polymorphism in these enzymes also plays a role in the susceptibility to adverse health effects from exposure to environmental pollutants (Uuskala et al., 1995).

The aim of this study is the evaluation of damage induced in humans exposed to different pesticide mixtures in Santa Fe province, Argentina, in order to estimate possible toxicity mechanisms and its relationship with labor features. The assessment of damage generated by direct exposure to pesticide mixtures was carried out through oxidative stress, DNA damage and susceptibility biomarkers in peripheral blood.

Study Population

The Regional Ethical Committee established the regulations for the development of the study and informed consent was given by each individual prior to the beginning of the study. A face-to-face questionnaire was completed to obtain information on a) standard demographic data (age, gender, among others), b) individual lifestyle (diet, smoking habit, alcohol and medicine consumption), c) occupational aspects (working hours/days, years of exposure to pesticides, use of protective measures, among others), and d) pesticides used.

Up to the moment, the study involved 30 subjects from San Cristóbal department (Santa Fe Province, Argentina; Figure 2) divided into two groups. The first group consist of pesticide sprayers or pesticide applicators who work in extensive crop (cereals and oilseeds), and the second is the control group, constituted by donors from the same area, with no history of occupational exposure to pesticides or any potential genotoxic agent. Peripheral blood heparinized samples were obtained from all the subjects involved in the study.

Catalase Activity in Erythrocytes (CAT)

Erythrocytes were hemolyzed by adding ice-cold demineralized ultrapure water at a 1:100 dilution. CAT activity in hemolyzed erythrocytes was measured spectrophotometrically by monitoring the decrease in H_2O_2 concentration over time (Aebi, 1984). The specific activity of each sample was calculated on the basis that one unit of enzyme activity was defined as the activity required to degrade 1 mole hydrogen peroxide during 60 s/g Hb.

Lipid Peroxidation in Erythrocytes (TBARS)

Malondialdehyde (MDA) as a marker of lipid peroxidation in red blood cells (dilution 1:4 with ice-cold) was determined by measuring the formation of the color produced during the reaction of thiobarbituric acid (TBA) with MDA (TBARS Assay) according to a modification of the method of Beuge & Aust (1978). The sample absorbance was determined at 535 nm and the TBARS concentration was calculated using the extinction coefficient 1.56 x 105 M-1 cm-1. MDA concentration in erythrocytes was expressed as nmol/g Hb.

Micronucleus Test in Buccal Epithelial Cells (BCMN)

Buccal cell suspensions were washed three times with a buffer solution followed by centrifugation at 1500 rpm, in order to remove bacteria and cell debris, which can confound the scoring. Smears were made and then slides were fixed in 80% cold methanol for 30 min, air-dried overnight (Titenko-Holland et al., 1998) and stored at $-20^{\circ}C$ until analysis. They were then stained with acridine orange (DNA specific stain) and analyzed using a fluorescence microscope at 400× magnification. A total of 2000 cells/subject were scored. The frequency of cells with MN per 1000 buccal cells was determined for each volunteer.

FPG-Modified Comet Assay

A freshly prepared suspension of 50 µl blood in 950 µl of RPMI 1640 at 37 °C was centrifuged at 1000*g* for 4 min. Cell pellets from volunteers were embedded in agarose, lysed, incubated with FPG or with enzyme buffer alone for 30 min at 37 °C, and then subjected to alkaline incubation and electrophoresis as in the standard assay (Singh et al., 1988). The level of DNA breaks induced by oxidative damage was calculated by subtracting breaks with buffer from breaks with FPG. Net FPG-sensitive sites (FPG sites) are a measure of oxidized purines (Collins et al., 2008). FPG sites = Damage Index CA with FPG – Damage Index CA without FPG.

DNA Extraction and Genotyping

Genomic DNA was obtained from 250 µl of whole blood using a commercially available kit according to the manufacturer instructions (NucleoSpin® Tissue, Mancherey-Nagel).

GSTM1 and GSTT1 genes were typed by a multiplex PCR method (Abdel-Rahman et al., 1996) with modifications. Reaction mixture consisted of 150 ng of genomic DNA, 15 pmol of each primer, 100mM Buffer PCR 10x, 1.5 mM $MgCl_2$, 200 pmol dNTPs and 2 U Taq DNA polymerase in a total volume of 50 µl. An aliquot of the amplification product was electrophoresed in horizontal agarose gel (2.5%) to verify the presence or the absence of GSTM and GSTT fragments and the CYP1A1 product was used as control for this reaction. The sequences of CYP1A1 primers were 5'-GAA CTG CCA CTT CAG C TG TCT and 5'-CAG CTG CAT TTG GAA GTG CTC; the sequences for GSTT1 primers were 5' GAA CTC CCT GAA AAG CTA AAG C and 5' GTT GGG CTC AAA TAT ACG GTG G, and finally for GSTM1 primers were 5' TTC CTT ACT GGT CCT CAC ATC TC and 5'TCA CCG GAT CAT GGC CAG CA.

Statistical Analysis

Data analysis was accomplished by using the SPSS 14.0 for Windows (2005) program. The normality of all variables was tested by Kolmogorov-Smirnov test, and homogeneity of variances by Levene test.

Non-parametric tests Kruskal-Wallis and Mann-Whitney were used to compare different groups, according to the different analysed parameters (biomarkers of effect and susceptibility), confusion factors (sex, age, habits, residence time), or occupational parameters (tasks, years of work, protective equipment, type of crop). A significant level was considered at $p < 0.05$.

Results

The average age group of workers was 40.66 ± 15.17 years and the control group was 37.70 ± 14.07 years, ranging between 20 and 55 years of age for all volunteers included. There were no differences in age, smoking habit, or alcohol consumption between the two groups ($p > 0.05$ in all analysis performed).

The analysis of data related to personal protecting equipment (PPE) showed that 62.5% of the workers do not wear any kind of protection, and only 16.5% of them wear at least 2 elements of protection.

Below is the list of pesticides most commonly used by the workers in the week prior to sampling (questionnaire answers).

Fungicide:

- Mancozeb (Dithiocarbamate-zinc).

Insecticides:

- Chlorpyrifos (Organophosphorus),
- Cypermethrin (Pyrethroid),
- Dimethoate (Organophosphorus),
- Endosulfan (Organochlorine),
- Pirimicarb (Metilcarbamate),
- Methamidophos (Organophosphorus),
- Monocrotophos (Organophosphorus).

Herbicides:

- Glyphosate (Phosphonoglycine),
- Flumetsulam (Triazolo Pyrimidine),
- 2,4D (Clorofenoxid acid),
- Paraquat (Bipiridilium),
- Picloram (Piridincarboxylic Acid).

In Table 1 we present results from oxidative status and genotoxicity biomarkers. We observed a significantly increased in TBARS levels (47%), CAT (61%), FPG sites (62 %) and BCMN (98 %) in pesticide applicators compared to the control group ($p<0.01$ in all analysis performed).

When analysing confounding factors in relation to biomarkers in applicators, no statistically significant differences ($p> 0.05$) were found for CAT, TBARS and FPG sites in respect to age, smoking and alcohol consumption, while smoking and alcohol consumption significantly affected BCMN results ($p< 0.05$). In the same way, there were differences in FPG sites and BCMN in relation to the factor "years of work" ($P < 0.05$).

Table 1. Mean values and standard deviation of the oxidative status and genotoxicity biomarkers in controls and applicators in extensive crops

Biomarkers of effect (X±DE)	Controls	Pesticide applicators
CAT (kU g^{-1} Hb)	169.09 ± 86.32	272.19 ± 126.71*
TBARS (nmol g^{-1} Hb)	191.33 ± 32.99	281.69 ± 66.09*
FPG Sites (arbitrary units)	14.28 ± 5.66	23.14 ± 9.95*
BCMN	0.59 ± 0.36	1.17 ± 0.69*

* $p<0.01$, Mann Whitney test.

The distribution of GSTM1 and GSTT1 genotypes showed that 87.5% of workers were GSTT1 positive, while 54.16% were GSTM1 positive, both under the Hardy-Weinberg balance. There was a significant increases in FPG sites ($p< 0.05$) in applicators with null GSTT1 polymorphism compared to positive ones, whereas no difference was observed between null and positive GSTM1 polymorphism in any biomarker ($p > 0.05$).

Discussion

All agricultural workers included in this study used more than two different pesticides and many of them were exposed to a great number of compounds. Some of these pesticides are classified as carcinogenic by the US Environmental Protection Agency (US-EPA) and hazardous by the World Health Organization (WHO), although not yet listed by the IARC.

The association of the effects observed with exposure to a single pesticide is very difficult because most occupations involve the regular use of a large number of different pesticides, together with other chemicals such as co-formulants, which vary greatly in their toxicities. It is known that, given all these possible variations in determinants of agricultural exposure (and pesticides in particular), it is difficult to accurately assess the intensity, duration, and frequency of exposures (Kromhout & Heederik, 2005) and to obtain correlations between increased genotoxicity biomarkers and the degree of exposure.

The proximity of housing to agricultural fields treated with pesticides has been suggested as a factor closely linked to exposure (Fenske et al., 2000). Studies of dust samples from farmworker households support this suggestion, both in terms of pesticide concentrations (McCauley et al., 2001) and the number of pesticides found in the home (Quandt et al., 2002, 2004). Different occupational situations, formulations used, mixtures applied, as well as agricultural practices in different regions, make it necessary a constant biomonitoring of the effects on human health.

Oxidative damage is thought to be an important mechanism of several pesticides (Banerjee et al., 1999, Prakasam et al., 2001). In blood, normal erythrocyte function depends on intact cell membrane, which is the target for many toxics, including pesticides. The results of the present study indicate that CAT activity increased significantly in extensive pesticide applicators. The available data on experimental animals (Seth et al., 2001), *in vitro* studies (Gultekin et al., 2000; Prasanthi et al., 2005) and *in vivo* studies (Ranjbar et al., 2002; Lopez et al., 2007) indicate that the enzymes associated with the antioxidant defence mechanism change under the influence of pesticides. These enzymes efficiently scavenge toxic free radicals and are partly responsible for protection against lipid peroxidation due to pesticide exposure (Banerjee et al., 1999). Hence, the increased level of TBARS observed in this work indicated peroxidation of membranes caused by exposure to pesticide mixtures, even though antioxidant activity showed and increment.

DNA oxidized basis can provide a useful marker in studies in which oxidative disturbances and genotoxicity occur, as an indicator of the general redox state of the organism. The combination of enzymes with a method that detects DNA breaks, such as Comet Assay, provides an alternative approach (Dusinska & Collins, 2008). This modification increases the sensitivity of the assay, since DNA oxidative damage is related to several environmental exposures and could be related to different pathophysiological factors. Muniz et al. (2008) showed that a significant amount of DNA damage induced by pesticides in applicators from United States could be caused by oxidative stress. Similar results were obtained by Gabbianelli et al. (2009) in rats exposed *in vivo* to permethrin through the application of the Comet Assay in combination with the FPG enzyme.

In the past few years, special attention has been paid to identifying inter-individual differences cause by a particular susceptibility to xenobiotics in some subjects. It has been proved that different polymorphic genes are responsible for the metabolism of some pesticides. The GSTs are a group of polymorphic enzymes involved in the detoxification of electrophilic metabolites generated in phase I of metabolism, and which also protect the organism from oxidative damage (Berhane et al., 1994). Null GSTM1 and GSTT1 genotypes have been associated to an increase in the frequency of cancer and cytogenetic damage (Pastor et al., 2003). In the present study, there were not differences in BCMN or FPG sites between positive and null GSTM1, similar to previous investigations (Da Silva et al., 2008; Liu et al., 2006). However, a significant increase in FPG sites in applicators with null GSTT1 was observed. Other researchers obtained increases on damage both for null GSTT1 (Abhishek et al., 2010) and GSTM1 (Singh et al. 2010, 2011) polymorphisms. Although available data about agricultural populations suggest that people with unfavourable alleles in their metabolic genes are more susceptible to genotoxic effects, there is not any conclusive result on whether these metabolic polymorphisms affect chromosomal damage induced by pesticides (Bolognesi et al., 2011). In Argentina, GSTT1 and GSTM1 polymorphisms have been only characterized regarding the prevalence of certain kind of cancer in a monitoring study in Cordoba province (Moore et al., 2004). The distribution of these genes tested in 215 people agreed with the ones observed in this study. These results encourage the use of GSTT1 and GSTM1 polymorphisms as biomarkers of susceptibility for further evaluations of environmentally or occupationally pesticide exposure.

6. Pesticides and Effects in Wildlife

Current agricultural related practices affect natural ecosystems through their conversion into agricultural lands and through the utilization of agrochemicals, which subsequent spread and runoff contaminate surrounding natural habitats (Johanson, 2004; Peruzzo et al., 2008). Agriculture expansion has led to increased fragmentation of habitat due to deforestation and high degradation of remnant ecosystems, with deep consequences for biodiversity. A great amount of pesticides applied in agriculture disperse through the environment, affecting wild flora and fauna populations of surrounding natural areas (Donald, 2004). Chronic exposure to chemicals, even at very low concentrations, may interfere with development and growth, hematological and physiological parameters and genetic stability of organisms living there (Glusczak et al., 2007).

It has been indicated that one of the main mechanisms of pesticide toxicity is ROS production and the corresponding oxidative damage may be one of the major causes of impairments in organisms. Studies showed that Roundup® (glyphosate) exposure may induce oxidative damage at different tissues leading to the generation of free radicals, and lipid peroxidation could be a mechanism involve in its toxicity. This was demonstrated for different species of fish such as *Rhamdia quelen, Prochilodus lineatus*, and *Cyprinus carpio* (Glusczak et al., 2007; Modesto & Martinez 2010; Cattaneo et al., 2011). Rossi et al. (2011) reported a reduced in CAT activity and increase in hidroperoxides in the fish *Astianax* caused by the herbicide Roundup® alone and also in mixture with another herbicide (diuron). In the Bullfrog (*Lithobates catesbeiana*) exposed to Roundup Original® (glyphosate 41%) the activities of the antioxidant enzymes SOD and CAT increased in the liver and decreased in muscle, while there was an increase in lipid peroxidation, showing that both the generation of reactive oxygen species and oxidative stress are involved in the toxicity induced by Roundup (Costa et al., 2008). Ballesteros et al. (2009) reported an increase in the levels of lipid peroxidation in the fish *Jenynsia multidentata* exposed to endosulfan.

Some authors suggest that ectothermic vertebrates have a low antioxidant capacity that would lead to an unfavorable balance between prooxidants and antioxidants (Voituron et al., 2006).

Biomonitoring of Reptile Populations Exposed to Pesticide in Santa Fe Province (Argentina)

Environmental biomonitoring make use of 'sentinel' organisms living in their natural habitat and reflecting long-term, continuous exposure. In north-central region of Argentina, many natural environments where the broad-snouted caiman (*Caiman latirostris*) live are in the proximity of regions with high agricultural activity, where glyphosate, endosulfan and cypermethrin are extensively used as a method for pest control in soybean crops. In addition, they are extensively applied in the same period of the year in which *C. latirostris* breeding season takes place, implying a contamination risk particularly important for developing embryos and neonates, which remain in small water bodies in surroundings (Poletta et al., 2009). Different studies in Argentina reported residues of pesticides in surface water and lagoons (Menone et al., 2001; Peruzzo et al., 2008).

In previous studies under laboratory controlled conditions, we demonstrated genotoxic effect through the Micronucleus (MN) test and Comet assay (CA) in *C. latirostris* neonates after *in ovo* exposure to increasing concentrations of the Glyphosate-based formulation Roundup® (RU; 66.2 % glyphosate), applied directly on the eggshell (topication) (Poletta et al., 2009). Similar results were obtained in neonates maintained during two months in water (in a semi-immersion system) with concentration of RU similar to those applied at crops. Then, we conducted field-like experiments simulating the exposure that a caiman nest can receive in neighboring croplands habitats, using pesticide practices commonly applied in agriculture. We found biochemical alterations, genotoxicity and effect on growth in caimans exposed to the mixture of glyphosate (66.2 %), endosulfan (35 %) and cypermethrin (25 %) formulations, as well as in those exposed to the single formulation Roundup® (Poletta et al., 2011). These results showed that the extensive use of pesticides on crops may cause environmental problems in wildlife of surrounding habitat.

Considering our previous results in genotoxicity induced by these pesticides in *C. latirostris* and information reported by other authors about oxidative stress and alterations in antioxidant defense systems in different species, we postulate that these parameters would be good early markers of pesticide effects in wild populations of reptile species.

Measuring DNA damage and repair with the Comet assay is commonly applied in laboratory or semi-natural conditions in different species, but, as we have seen, the specificity of the assay is greatly improved if lesion-specific enzymes are included. There are still only a limited number of reports of the use of the CA in an ecotoxicological context, and very few of these use lesion-specific enzymes to detect specifically oxidised bases (Azqueta et al., 2009). Up to our knowledge no studies have been made that evaluate DNA oxidative damage through the modified comet assay on reptile species.

Study Population

The study takes place in two different areas situated in San Cristóbal department, Santa Fe province (Argentina) (Figure 2):

1. area exposed to pesticides, situated near to agricultural activities center: "El Lucero" (29° 54' S; 50° 21'W).
2. area not exposed to pesticides (natural reserve, law N 12930, Santa Fe, Argentina), free of any contamination activity: "El Fisco" (30° 15' S; 60° 50' W). This is the control area.

All animals are treated following the Ethical Reference Framework for Biomedics Researches: ethical principles for research with laboratory, farm and wild animals (CONICET, 2005), using non-harmful techniques of blood collection and minimizing stress and suffering by suitable management methods. Samples from caimans of different size (from hatchlings to adults) are taken from the spinal vein (Zippel et al., 2003) with heparinised disposable syringes.

Modifications in Protocols of Oxidative Damage and Antioxidant Defense Capacity Markers to Be Applied in C. latirostris

There was no previous data on the application of any oxidative stress technique in *C. latirostris* and only a few reports were found in other reptile species as indicators of natural stressors like extreme cold (Valdivia et al., 2006; Voituron et al., 2006). In crocodilians, studies reported basal values of lipid peroxidation in *Caiman yacare* (Furtado Filho et al, 2007) or modifications in the level of TBARS and GST in *Alligator mississipiensis* from contaminated environments related to heavy metals (Gunderson et al., 2004; Lance et al., 2006). All these studies were applied in tissues from kidney, muscles, gonads or liver, so that animals have to be sacrificed or samples obtained from animals recently dead.

In our study, different modifications are being done to standard procedures in order to adapt the techniques for their application in blood of *C. latirostris*. In this way, they can be used as biomarkers of oxidative stress produced by pesticides in wild populations, without causing any damage to the animals.

FPG-Modified Comet Assay

The alkaline Comet assay is performed with modifications required by *C. latirostris* erythrocytes, as described in Poletta et al. (2008). Blood samples are diluted 1:19 (v/v) with RPMI-1640, 1.5 µl of each diluted sample (4.0×10^3 erythrocytes, approximately) is added to 100 ml of LMP agarose and slides prepared following standard procedure. After lysis, slides are incubated with FPG or with enzyme buffer alone for 30 min at 37 °C, submerged in alkaline buffer for 10 min and then electrophoresed at 0.90 V/cm during 10 min too. One hundred randomly selected comet images are analyzed, classified into five arbitrary classes, and a single DNA damage index (DI = $n1+2\ n2+3\ n3+4\ n4$) calculated for each animal

(Poletta et al., 2008). DNA breaks induced by oxidative damage are calculated by subtracting breaks with buffer from breaks with FPG as follows: FPG sites = Damage Index CA with FPG – Damage Index CA without FPG (Collins et al., 2008).

It is important to highlight that this is the first report on the application of the FPG-modified Comet Assay in a reptile species.

DNA Repair Assay

We evaluate *C. latirostris* erythrocytes resistance to oxidative DNA damage induced by hydrogen peroxide (H_2O_2). Briefly, 50 µl whole blood are suspended in 925 µl RPMI 1640 medium. Cell dilution is mixed with 25 µl of H_2O_2 solution (10 mM) and allowed to damage the DNA for 20 min at 25 °C. The reaction is quenched using 500 µl of 2% DMSO solution in PBS. Each sample is centrifuged at 1000*g*, washed again, and suspended in RPMI 1640 medium supplemented with 10% fetal bovine serum. We test time necessary for repairing in this species, incubating samples for 30 min, 1, 2, 12 and 24 h at 25 °C. Immediately after incubation time, we prepare the slides following the standard procedure for comet assay application in *C. latirostris* (Poletta et al., 2008) and Damage index of repair (DIR) is calculated.

MN Test in Erythrocytes

The MN assay is applied in *C. latirostris* erythrocytes as previously described (Poletta et al., 2008). Smears are prepared, coded for 'blind' analysis and stained with acridine orange supravital stain. In each sample, 1000 erythrocytes are analyzed and the frequency of MN is determined as follows: FMN= N° erythrocytes with MN/1000erythrocytes scored.

Catalase (CAT) Activity in Erythrocytes

Erythrocytes are hemolyzed by adding ice-cold demineralized ultrapure water at a 1:100 dilution. CAT activity in hemolyzed erythrocytes is measured spectrophotometrically by monitoring the decrease in H_2O_2 concentration over time (Aebi, 1984). The specific activity of each sample is calculated on the basis that one unit of enzyme activity was defined as the activity required to degrade 1 mole hydrogen peroxide during 60 s/g Hb.

Lipid Peroxidation in Erythrocytes (TBARS)

Malondialdehyde (MDA) as a marker of lipid peroxidation in red blood cells (dilution 1:20 with ice-cold) is determined by measuring the formation of the color produced during the reaction of thiobarbituric acid (TBA) with MDA (TBARS Assay) according to a modification of the method of Beuge & Aust (1978). The sample absorbance is determined at 535 nm and the TBARS concentration is calculated using the extinction coefficient 1.56 x 105 M-1 cm-1. MDA concentration in erythrocytes is expressed as nmol/g Hb. Initially, different dilutions of the red blood cells with ice-cold were tried: 1:10, 1:20, 1:40. We observed that the proper dilution for the determination of TBARS in *C. latirostris* red blood cells is 1:20, so this has been the one used since then.

Reduced and Oxidazed Glutathione Relation (GSH/GSSG)

When ROS increase, GSSG formation exceeds elimination and the relation GSH/GSSG decreases, being a useful indicator of oxidative stress. Erythrocytes are lised by adding ice-cold water and then mixed with trichloroacetic acid. GSH is determined in the supernatant following the procedure described by Ellman (1961). For the determination of total glutathione (GSH + GSSG), GSSG is reduced to GSH with glutathione reductase and NADPH during appropriate time at room temperature. Reaction is stopped by acid precipitation with trichloroacetic acid (Venturino et al., 2001). Dithionitrobenzoic acid (DTNB) is added to the supernatant and absorbance read at 412 nm. Total glutathione is expressed as μM of glutathione mg^{-1} Hb and then the relation GSH/GSSG is calculated.

Superoxide Dismutase (SOD) Activity in Erythrocytes

SOD activity is determined by quantifying the inhibition of the reduction of nitroblue tetrazolium (NBT) by the $O_2^{\cdot-}$ -generated as a by-product of the xanthine/xanthine oxidase system at 560 nm, as described by Valdivia et al. (2006). In *C. latirostris* the assay is applied on blood and carried out at 29-30 °C. One unit of SOD activity is defined as the amount of enzyme needed to inhibit 50% of the $O_2^{\cdot-}$ - NBT reaction.

Final Comments

Up to the moment, we have adapted the techniques described obtaining good results for their application in *C. latirostris* blood. This means an important advance for the evaluation of wild caiman populations, as we obtain samples without causing any damage to the animals. All these biomarkers are currently being used for the evaluation of pesticides effects on caimans environmentally exposed in Argentina. An integrative approach including the analysis of different biomarkers with multiple endpoints like oxidative stress, genotoxicity and physiological parameters could be highly informative for a better understanding of the mechanisms underlying pesticide damage, providing predictive tools for the environmental protection of this and other related species.

Acknowledgment

This work was supported by Universidad Nacional del Litoral (PI 09-50), Consejo Nacional de Investigaciones Científicas y Técnicas– CONICET (PIP 5012 to MDM), Proyecto Yacaré and Yacarés Santafesinos (Gob. Sta. Fe/MUPCN), and Universidad de Buenos Aires (UBACyT X154 to MDM). We are especially grateful to Dr. Andrew R. Collins (Department of Nutrition, Institute of Basic Medical Sciences, University of Oslo) for providing FPG enzyme.

References

Abdel-Rahman, S. Z.; El-Zein, R. A.; Anwar, W. A.; Au, W.W. 1996. A multiplex PCR procedure for polymorphic analisys of GSTM1 and GSTT1 genes in population studies. *Cancer Lett.* 107: 229-233.

Abhishek, S.; Kaur, N.; Kaur, S.; Lata, M.; Sarma, J.K.; Shama, A. 2010. Association of GSTM1 and GSTT1 Gene Deletions with Susceptibility to DNA Damage in the Pesticide-Exposed Workers of Punjab. *Rejuv. Res.* 13: 281-287.

Aebi, H. (1984). Catalase in vitro, *Method. Enzymol.*, 105, 121-6.

Aizen, M.; Garibaldi, L. A.; Dondo, M. 2009. Expansión de la soja y diversidad de la agricultura argentina. *Ecol. Austral* 19: 45-54.

Alavanja, M. C. R.; Hoppin, J. A.; Kamel, F. 2004. Health Effects of Chronic Pesticide Exposure: Cancer and Neurotoxicity. *Annu. Rev. Publ. Health*. 25: 155-197.

Amr, M. M., 1999. Pesticide monitoring and its health problems in Egypt; a Third World country. *Toxicol. Lett.* 107: 1-13.

Azqueta, A.; Shaposhnikov, S.; Collins, A. R. 2009. DNA oxidation: Investigating its key role in environmental mutagenesis with the comet assay. *Mutat. Res.* 674: 101–108.

Ballesteros, M. L.; Wunderlin, D. A.; Bistoni, M. A. 2009. Oxidative stress responses in different organs of *Jenynsia multidentata* exposed to endosulfan. *Ecotox. Environm. Saf.* 72 (1): 199-205.

Banerjee, B. D.; Seth, V.; Bhattacharya, A.; Pasha, S. T.; Chakraborty, A. K. 1999. Biochemical effects of some pesticides on lipid peroxidation and free-radical scavengers. *Toxicol. Lett.*107: 33-47.

Bassil, K. L.;Vakil, C.; Sanborn, M.; Cole, C.; Kaur, J. S.; Kerr, K. J. 2007. Cancer health effects of pesticides: systematic review. *Can. Fam. Physician* 53: 1704-1711.

Berhane, K.; Widersten, M.; Engström, A.; Kozarich, J.W.; Mannervik, B. 1994. *Detoxification of base propenals and other α;β-unsaturated aldehyde products of radical reactions and lipid peroxidation by human glutathione transferases*. Proc. Natl. Acad. Sci. USA. 91: 1498-1484.

Beuge, J. A.; Aust, S. D. 1978. Microsomal lipid peroxidation. *Method. Enzymol.* 52: 302-10.

Bickham, J.W.; Sandhu, S.; Hebert, P.D.N.; Chikhi, L.; Athwal, R. 2000. Effects of chemical contaminants on genetic diversity in natural populations: implications for biomonitoring and ecotoxicology. *Mutat. Res.* 463: 33-51.

Board, P.; Coggan, M.; Johnson, P.; Ross, V.; Suzuki, T. 1990. Genetic heterogeneity of the human glutathione S-Transferases: a complex of gene families. *Pharmacol. Ther.* 48: 357-369.

Bolognesi, C; Parrini, M.; Bonassi, S.; Ianello, G.; Salanitto, A. 1993. Cytogenetic analysis of a human population occupationally exposed to pesticides. *Mutat. Res.* 285: 239-249.

Bolognesi, C.; Creus, A.; Ostrosky-Wegman, P.; Marcos, R., 2011. Micronuclei and pesticide exposure. *Mutagenesis* 26 (1): 19-26.

Carbonell, E.; Puig, F.; Xamena, N.; Creus, A.; Marcos, R. 1990. Sister chromatid exchange in lymphocytes of agricultural workers exposed to pesticides. *Mutagenesis* 5: 403-405.

CASAFE. *Cámara de Sanidad Agropecuaria y Fertilizantes Argentina*. Available on: http://www.casafe.org/ (12/15/2011).

Cattaneo, R.; Clasen, B.; Loro, V.L.; Cavalheiro de Menezes, C.; Pretto, A. Baldisserotto, B.; Santi, A.; Antonio de Avila, L. 2011. Toxicological Responses of *Cyprinus carpio* Exposed to a Commercial Formulation Containing Glyphosate. *Bull. Environ. Contam.Toxicol.* 87: 597-602.

Çok, İ.; O. K. Ulutaş, Ö. Okuşluk, E. Durmaz, N. Demir. 2011. Evaluation of DNA Damage in Common Carp (Cyprinus carpio L.) by Comet Assay for Determination of Possible Pollution in Lake Mogan (Ankara). *Scient. World J.* 11: 1455–1461.

Collins, A. R. 2004. The Comet Assay for DNA Damage and Repair, Principles, Applications, and Limitations. *Mol. Biotech* 26: 249-261.

Collins, A. R. 2009. Investigating oxidative DNA damage and its repair using the comet assay. *Mutat Res* 681: 24-32.

Collins, A. R.; Oscoz, A. A.; Brunborg, G.; Gaivao, I.; Giovannelli, L.; Kruszewski, M.; Smith, C. C.; Stetina, R. 2008. The comet assay: topical issues. *Mutagenesis* 23 (3): 143-151.

CONICET. Consejo Nacional de Investigaciones Científicas y Técnicas 2005. *Ethical Reference Framework for Biomedics Researches: ethical principles for research with laboratory, farm and wild animals.* Available from: /http://www.conicet. gov.ar/webfiles/ 2010/01/OCRRD20050701-1047.pdfS.

Cornetta, T.; Padua, L.; Testa, A.; Levoli, E.; Festa, F.; Tranfo, G.; Baccelliere, L.; Cozzi, R. 2008. Molecular biomonitoring of a population of nurses handling antineoplastic drugs. *Mutat. Res.* 638: 75-82.

Costa, M. J.; Monteiro, D. A.; Oliveira-Neto, A. L.; Rantin, F. T.; Kalinin, A. L. 2008. Oxidative stress biomarkers and heart function in bullfrog tadpoles exposed to Roundup Original®. *Ecotoxicol.* 17:153–163.

Da Silva, J.; Moraes, C. R.; Heuser, V. D.; Andrade, V. M.; Silva, F. R.; Kvitko, K.; Emmel, V.; Rohr, P.; Bordin, D. L.; Andreazza, A. C.; Salvador, M.; Henriques, J. A. P. Erdtmann, B. 2008. Evaluation of genetic damage in a Brazilian population occupationally exposed to pesticides and its correlation with polymorphisms in metabolizing genes. *Mutagenesis* 23 (5): 415-422.

Dhawan, A.; Bajpayee, M.; Parmar, D. 2009. Comet assay: a reliable tool for the assessment of DNA damage in different models. *Cell Biol. Toxicol.* 25: 5–32.

De Andrade, V. M., de Freitas, T. R., da Silva, J. 2004. Comet assay using mullet (Mugil sp.) and sea catfish (Netuma sp.) erythrocytes for the detection of genotoxic pollutants in aquatic environment. *Mutat. Res.* 560: 57-67.

Donald, P. F. 2004. Biodiversity impacts of some agricultural commodity production systems. *Conserv. Biol.* 18: 17-37.

Dulout, F. N.; Lopez Camelo, J. S.; Guradze, H. N. 1992. Analysis of sister chromatid exchanges (SCE) in human population studies. *Rev. Brasil. Genet.* 15: 169-182.

Dusinska, M.; Collins, A. R. 2008. *The Comet assay in human biomonitoring.*

Ellman, G. L.; Curtney, K. D.; Andrews, V.; Featherstone, R. M. 1961. A new and rapid colorimetric determination of acetyl cholinesterase activity. *Biochem.Pharmacol.* 7: 88–95.

Falck, G. C. M.; Hirvonen, A.; Scarpato, R.; Saarikoski, S. T.; Migliore, L. y Norppa, H. 1999. Micronuclei in blood lymphocytes and genetic polymorphism for GSTM1; GSTT1 and NAT2 in pesticide-exposed greenhouse workers. *Mutat. Res.* 441: 225-237.

Fenech, M. 2000. The *in vitro* micronucleus technique. *Mutat. Res.* 455: 81-95.

Fenske, R. A.; Kissel, J. C.; Lu, C.; Kalman, D. A.; Simcox, N. J.; Allen, E. H. 2000. Biologically based pesticide dose estimates for children in an agricultural community. *Environ. Health Perspect.* 108: 515-520.

Franco, R.; Sánchez-Olea, R.; Reyes-Reyes, E. M.; Panayiotidis, M. I. 2009. Environmental toxicity, oxidative stress and apoptosis: Ménage à Trois. *Mutat. Res.* 674: 3-22.

Frenzilli, G.; Nigro, M.; Lyons, B. P. 2009. The Comet assay for the evaluation of genotoxic impact in aquatic environments. *Mutat. Res.* 681: 80–92.

Furtado-Filho, O. V.; Polcheira, C.; Machado, D. P.; Mourão, G.; Hermes-Lima, M. 2007. Selected oxidative stress markers in a South American crocodilian species. *Comp. Biochem. Physiol.* C 146 (1-2): 241-54.

Gabbianelli, R.; Falcioni, M. L.; Cantalamessa, F.; Nasuti, C. 2009. Permethrin induces lymphocyte DNA lesions at both Endo III and Fpg sites and changes in monocyte respiratory burst in rats. *J. Appl. Toxicol.* 29: 317-322.

Glusczak, L.; Miron, D. S.; Moraes, B. S.; Simoes, R. R.; Schetinger, M. R. C.; Morsch, V. M.; Loro, V. L. 2007. Acute effects of glyphosate herbicide on metabolic and enzymatic parameters of silver catfish (*Rhamdia quelen*). *Comp. Biochem. Physiol.* C 146: 519-524.

Goetz, M. E.; Luch, A. 2008. Reactive species: A cell damaging rout assisting to chemical carcinogens. *Cancer Lett.* 266, 73–83.

Gomez-Arroyo, S.; Diaz-Sanchez, Y.; Meneses-Perez, M.A.; Villalobos-Pietrini, R., De Leon-Rodriguez, J. 2000. Cytogenetic biomonitoring in a Mexican floriculture worker group exposed to pesticides. *Mutat. Res.* 466: 117-124.

Grisolia, C. K. 2005. *Agrotóxicos: Mutações, Câncer e Reprodução.* Editora da UNB, Brasilia, 392 pp.

Gultekin, F.; Ozturk, M.; Akdogan, M. 2000. The effect of organophosphate insecticide chlorpyrifos-ethyl on lipid peroxidation and antioxidant enzymes (*in vitro*) *Arch. Toxicol.* 74 (9): 533-538.

Gunderson, E.; Oberdörster, A.; Guillette, L. J. Jr. 2004. Phase I and II liver enzyme activities in juvenile alligators (*Alligator mississippiensis*) collected from three sites in the Kissimmee–Everglades drainage, Florida (USA). M.P. *Comp. Biochem. Physiol.* C 139 (1-3): 39-46.

Hang, S., Rampoldi, E. A., Negro, G. J. *Herbicide Behaviour in Non Tillage Systems*, *in*Piotrowski, K.D. (Ed.), Herbicides: Properties, Crop Protection and Environmental Hazards, Chapter 3, Nova Science Publishers, New York, USA, 2011. In press.

Jha, A. N. 2008. Ecotoxicological applications and significance of the comet assay. *Mutag.* 23 (3): 207-221.

Johanson, M., 2004. Effect of Agriculture on Abundance, Genetic Diversity and Fitness in the Common Frog, Rana temporaria. *Acta Univ. Upsal.*1025, 43.

Kromhout, H.; Heederik, D. 2005. Effects of errors in the measurement of agricultural exposures. *Scand. J. Work. Environ. Health.* 31: 33-38.

Lance, V. A.; Horn, T. R.; Elsey, R. M.; de Peyster, A. 2006. *Chronic incidental lead ingestion in a group of captive–reared alligators* (*Alligator mississippiensis*): possible contribution to reproductive failure. *Comp. Biochem. Physiol.* C 142: 30–35.

Lander, F.; Knudsen, L. E.; Gamborg, M. O.; Jarventaus, H.; Norppa, H. 2000. Chromosome aberrations in pesticide-exposed greenhouse workers. *Scand. J. Work. Environ. Health* 26: 436-442.

Lanteri, L. 2009. *Respuesta a precios del área sembrada de soja en la Argentina.* Documentos de Trabajo 2009, 46. Banco Central de la República Argentina *ie*, Investigaciones Económicas, ISSN 1850-3977, pp-1-39.

Lantieri, M. J., Meyer Paz, R., Butinof, M., Fernández, R. A., Stimolo, M. I.; Díaz, M. P. 2009. Exposición a plaguicidas en agroaplicadores terrestres de la provincia de Córdoba: Factores condicionantes, *Agriscientia* 26 (2): 43-54.

Limón-Pacheco, J.; Gonsebatt, M. E. 2009. The role of antioxidants and antioxidant-related enzymes in protective responses to environmentally induced oxidative stress. *Mini review. Mutat. Res.* 674: 137-147.

Liu, Y. J.; Huang, P. L.; Chang, Y. F; Chen, Y. H.; Chiuo, Y. H.; Xu, Z. L.; Wong R. H. 2006. GSTP1 Genetic Polymorphims is Associated with a Higher Risk of DNA Damage in Pesticide-Exposed Fruit Growers. *Cancer Epidemiol. Biom. Prev* 15 (4): 659-666.

Lopez, O.; Hernandez, A.F.; Rodrigo, L.; Gil, F.; Pena, G.; Serrano, J. L.; Parron, T.; Villanueva, E.; Pla, A. 2007. Changes in antioxidant enzymes in humans with long-term exposure to pesticides. *Toxicol.Lett.* 171: 146-153.

Mannervick, B.; Awasthi, Y. C.; Board, P. G.; Yayes, J. D.; Dillio, C.; Ketterer, B.; Litowski, I.; Morgenstem, R.; Muramatsu, M.; Pearson, W.R.; Pickett, C.B.; Sato, K.; Widerstem, M.; Wolf, C.R. 1992. Nomenclature for human glutathione transferase. *Biochem. J.* 282: 305-308.

Mañas, F.; Peralta, L.; Gorla, N.; Bosh, B.; Aissa, D. 2009. Aberraciones Cromosómicas en Trabajadores Rurales de la Provincia de Córdoba Expuestos a Plaguicidas. *J. Basic Appl. Genet.* 20 (1): 09-13.

McCauley, L. A.; Lasarev, M. R.; Higgins, G.; Rothlein, J.; Muniz, J.y Ebbert; C. 2001. Work characteristics and pesticide exposures among migrant agricultural families: a community-based research approach. *Environ. Health Perspect.* 109: 533-538.

McCauley, L. A.; Anger, W. K.; Keifer, M.; Langley, R.; Robson, M. G.; Rohlman, D. 2006. Studying health outcomes in farmworker populations exposed to pesticides. *Environ. Health Perspect.*114: 953-60.

McLaughlin, A.; Mineau, P. 1995. The impact of agricultural practices on biodiversity. *Agric. Ecos. Environ.* 55, 201-212.

Menone, M. L.; Aizpún de Moreno, J. E.; Moreno, V. J.; Lanfranchi, A. L.; Metcalfe T. L.; Metcalfe, C. D. 2001. Organochlorine Pesticides and PCBs in a Southern Atlantic Coastal Lagoon Watershed, Argentina.Arch. Environ. *Contam.Toxicol.* 40 (3): 355-62.

Mitchelmore, C. L.; Chipman, J. K. 1998. DNA strand breakage in aquatic organisms and the potential value of the comet assay in environmental monitoring. *Mutat. Res.* 399: 135-147.

Modesto, K. A.; Martinez, C. B. R. 2010. Roundup® causes oxidative stress in liver and inhibits acetylcholinesterase in muscle and brain of the fish *Prochilodus lineatus. Chemosphere* 78: 294-299.

Moore, L. E.; Wiencke, J. K.; Bates, M. N.; Zheng, S.; Rey, O. A.; Smith, A. 2004. Investigation of genetic polymorphisms and smoking in a bladder cancer case–control study in Argentina. *Cancer Lett.* 211: 199-207.

Muniz, J. F.; McCauley, L.; Scherer, J.; Lasarev, M.; Koshy, M.; Kow, Y.W.; Nazar-Stewart, V.; Kisby, G.E. 2008. Biomarkers of oxidative stress and DNA damage in agricultural workers: A pilot study. *Toxicol. Appl. Pharmacol.* 227: 97-107.

Navarrete, D. M.; Gallopín, G. C.; Blanco, M.; Díaz-Zorita, M.; Ferraro, D. O.; Herzer, H.; Laterra, P.; Muermis, M. R.; Podestá, G. P.; Rabinovich, J.; Satorre, E. H.; Torres, F.; Viglizzo, E. F. 2009. Multi-casual and integrated assesment of sustainability: the case of agriculturalization in the Argentine Pampas. *Environ. Dev. Sustain.* 11: 621-638.

Paruelo, J. M.; Guerschman, J. P.; Piñeiro, G.; Jobbágy, E. G.; Verón, S. R.; Baldi, G.; Baeza, S. 2006. Cambios en el uso de la tierra en Argentina y Uruguay: marcos conceptuales para su análisis. *Agrociencia* 47: 47–61.

Pastor, S.; Creus, A.; Parrón, T.; Cebulska-Wasilewska, A.; Siffel, C.; Piperakis, S.; Marcos R. 2003. Biomonitoring of four European populations occupationally exposed to pesticides: use of micronuclei as biomarkers. *Mutagenesis* 18 (3): 249-258.

Penna, J.; Lema, D. 2002. *Adoption of Herbicide Resistant Soybeans in Argentina: an Economic Analysis*, IESR, INTA.

Peruzzo, P. J.; Porta, A. A.; Ronco, A. E. 2008. Levels of glyphosate in surface waters, sediments and soils associated with direct sowing soybean cultivation in north pampasic region of Argentina. *Environ. Pollut.* 156: 61-66.

Piperakis, S. M.; Kontogianni, K.; Piperakis, M. M.; Tsilimigaki, S. 2006. Effects of pesticides on occupationally exposed humans. *Scient. World J.* 6: 1211-1220.

Poletta, G. L.; Larriera, A.; Kleinsorge; E., Mudry, M. D. 2008. *Caiman latirostris* (broad-snouted caiman) as a sentinel organism for genotoxic monitoring: Basal values determination of Micronucleus and Comet assay. *Mutat. Res.* 650: 202-209.

Poletta, G. L.; Larriera, A.; Kleinsorge, E.; Mudry, M.D. 2009. Genotoxicity of the herbicide formulation Roundup® (glyphosate) in broad-snouted caiman (*Caiman latirostris*) evidenced by the Comet assay and the Micronucleus test. *Mutat. Res.* 672: 95-102.

Poletta, G. L.; Kleinsorge, E; Paonessa, A; Mudry, M. D.; Larriera, A; Siroski, P.A. 2011. Genetic, enzymatic and developmental alterations observed in *Caiman latirostris* exposed *in ovo* to pesticide formulations and mixtures in an experiment simulating environmental exposure. *Ecotoxicol.Environm. Saf.* 74: 852–859.

Prakasam, A. S.; Sethupathy, S.; Lalitha, S. 2001. Plasma and RBCs antioxidant status in occupational male pesticide sprayers, *Clin.Chim. Acta* 310: 107-112.

Prasanthi, K.; Muralidhara, A. Rajini, P.S. 2005. Morphological and biochemical perturbations in rat erythrocytes following in vitro exposure to Fenvalerate and its metabolite. *Toxicol.in Vitro* 19: 449-456.

Quandt, S. A.; Arcury, T. A.; Mellen, B. G.; Rao, P.; Camann, D.E. y Doran; A.M.; 2002. Pesticides in wipes from farmworker residences in North Carolina. In: *Proceedings of the 9th International Conference on Indoor Air Quality and Climate*; Monterrey; CA (Levin H; ed). Santa Cruz; CA: Indoor Air. 900-905.

Quandt, S. A.; Arcury, T. A.; Rao, P.; Snively, B. M.; Camann, D. E.; Doran, A. M. 2004. Agricultural and residential pesticides in wipe samples from farmworker family residences in North Carolina. *Environ. Health.Perspect.* 112: 382-387.

Ranjbar, A.; Pasalar, P.; Abdollahi, M. 2002. Induction of oxidative stress and acetylcholinesterase inhibition in organophosphorous pesticide manufacturing workers. *Hum. Exp. Toxicol.* 21: 179-82.

Remor, A. P.; Totti, C. C.; Moreira, D. A.; Dutra, G. P.; Heuser, V. D.; Boeira, J. M. 2009. Occupational exposure of farm workers to pesticides: biochemical parameters and evaluation of genotoxicity. *Environ. Int.* 35 (2): 273-278.

Rossi, S. C.; Dreyer da Silva, M.; Stein Piancini, L. D.; Oliveira Ribeiro, C. A.; Cestari, M.M.; Silva de Assis, H.C. 2011. Sublethal Effects of Waterborne Herbicides in Tropical Freshwater Fish. Bull. *Environ. Contam. Toxicol.* 87: 603-607.

SAGPyA. Secretaría de Agricultura, Ganadería, pesca y Alimentos, Ministerio de Agricultura, Ganadería y Pesca de la Nación (MAGyP). *Estimaciones agrícolas: mapa de principales cultivos*. Available on: http: //www.minagri.gob.ar/SAGPyA/agricultura/ cultivos_en_la_argentina (11/13/2011).

Saura-Mas, S.; Boone, M. D.; Bridges, C. M. 2002. Evaluation of direct effects of an insecticide on Gray Treefrogs: laboratory and fields trials. *J. Herpetol.* 36: 715-719.

Scarpato, R.; Migliore, L.; Angotzi, G.; Fedi, A.; Miligi, L.; Loprieno, N. 1996. Cytogentic monitoring of a group of Italian floriculturists. No evidence of DNA damage related to pesticide exposure. *Mutat. Res.* 367: 73-82.

Schmid, W. 1975. The micronucleus test. *Mutat. Res.* 31: 9-15.

Seth, V.; Banerjee, B. D.; Chakravorty, A.K. 2001. Lipid peroxidation, free radical scavenging enzymes, and glutathione redox system in blood of rats exposed to Propoxur, Pest. *Biochem. Phys.* 71: 133-139.

Shugart, L.R. 2000. DNA damage as a biomarker of exposure. *Ecotoxicol.* 9: 329.

SIIA. Sistema Integrado de Información Agropecuaria. Ministerio de Agricultura, Ganadería, Pesca y Alimentos de la República Argentina (MAGPyA). *Estimaciones Agrícolas. Superficies implantadas por provincia.* Available on: http://www.siia.gov.ar/index.php/series-por-tema/agricultura (1/10/2012).

Simoniello, M. F.; Kleinsorge, E. C.; Scagnetti, J. A.; Grigolato, R. A.; Poletta, G. L.; Carballo, M. A. 2008. DNA damage in workers occupationally exposed to pesticide mixtures. *J. Appl. Toxicol.* 28 (8): 957-965.

Simoniello, M. F.; Kleinsorge, E. C.; Scagnetti, J. A.; Mastandrea, C.; Grigolato, R.A.; Paonessa, A. M. Carballo, M. A. 2010. Biomarkers of cellular reaction to pesticide exposure in a rural population. *Biom,* 15 (1): 52-60.

Singh, S.; Kumar, V; Singh, P.; Dev Banerjee, B.; Singh Rautel, R.; Grover, S.S.; Singh Rawat, D.; Pash, S.T.; Jainc, S.K.; Rai, A. 2011. *Influence of CYP2C9, GSTM1, GSTT1 and NAT2 genetic polymorphisms on DNA damage in workers occupationally exposed to organophosphate pesticides*. Mutat.Res. In Press.

Singh, N. P.; McCoy, M. T.; Tice, R. R.; Schneider, E. L. 1988. A simple technique for quantitation of low levels of DNA damage in individuals cells. *Exp. Cell. Res.* 175 (1): 184-191.

Singh, S.; Kumar, V.; Singh P.; Thakur, S.; Banerjee, B.D.; Singh Rautela R.; Grover, S.S.; Singh Rawat, D.; Pasha, S.T.; Jain, S.K.; Rai, A. 2011. Genetic polymorphisms of GSTM1, GSTT1 and GSTP1 and susceptibility to DNA damage in workers occupationally exposed to organophosphate pesticides. *Mutat. Res.* 725: 36-42.

Stark, G. 2005. Functional consequences of oxidative membrane damage. *J. Membr. Biol.* 205: 1-16.

Storey, K. B. 1996. Oxidative stress: animal adaptations in nature. *Braz. J. Med. Biol. Res.* 29: 1715-1733.

Stratta Fernández, R. and Ríos Carmenado, I. 2010. Transformaciones agrícolas y respoblamiento en las comunidades rurales de la Región Pampeana Argentina. *Estudios Geográficos* 71 (268): 235-265.

Titenko-Holland, N.; Jacob, R. A., Shang, N.; Balaraman, A.; Smith, M. T. 1998. Micronuclei in lymphocytes and exfoliated buccal cells of postmenopausal women with dietary changes in folate. *Mutat. Res.* 417: 101-114.

Tsui, M. T. K; Chu, L. M. 2003. Aquatic toxicity of glyphosate-based formulations: comparison between different organisms and the effects of environmental factors. *Chemosphere* 52: 1189-1197.

Udroiu, I. 2006. The micronucleus test in piscine erythrocytes. *Aquat.Toxicol.* 79: 201-204.

Üner, N; Oruc, E. O.; Sevgiler, Y.; Sahin, N.; Durmaz, H.; Usta, D. 2006. Effects of diazinon on acetylcholinesterase activity and lipid peroxidation in the brain of *Oreochromis niloticus*.Environ. *Toxicol. Phar*,21:241-245.

Uuskala, M.; Jarventaus, H.; Hirvonen, A.; Sorsa, M.; Norppa, H. 1995. Influence of GSTMl genotype on sister chromatid exchange induction by styrene 7,8-oxide and 1,2-epoxy-3-butene in cultured human lymphocytes. *Carcinogenesis* 16: 947-950.

Valdivia, P. A.; Zenteno-Savín, T.; Gardner, S. C.; Aguirre, A. A. 2006. Basic oxidative stress metabolites in eastern Pacific green turtles (*Chelonia mydas agassizii*).*Comp. Biochem. Physiol,* C 146 (1-2): 111-117.

Venturino A.; Anguiano O.L.; Gauna L.; Cocca C.; Bergoc, R.M.; Pechen de D'Angelo A.M. 2001. Thiols and polyamines in the potentiation of malathion toxicity in larval stages of the toad *Bufo arenarum*. *Comp. Biochem. Physiol.* Part C 130: 191-198.

Voituron, Y.; Servais, S.; Romestaing, C.; Douki, T; Barré, H. 2006. Oxidative DNA damage and antioxidant defenses in the European common lizard (*Lacerta vivipara*) in supercooled and frozen states. *Cryobiol,* 52: 74-82.

World Health Organization (WHO). 1990. *Public health impact of pesticides used in agriculture*, Geneva, In: World Health Organization, 4.11.2010, Available from: http://whqlibdoc.who.int/publications/1990/9241561394. pdf.

Yáñez, L.; Ortiz, D.; Calderón, J.; Batres, L.; Carrizales, L.; Mejía, J.; Martínez, L.; García-Nieto, E.; Díaz-Barriga, F. 2002. Overview of Human Health and Chemical Mixtures: Problems Facing Developing Countries. *Environ, Health Perspect*, 110: 901-909.

Zippel, K.C.; Lillywhite, H.B.; Mladinich, C.R.J. 2003. Anatomy of the crocodilian spinal vein. *J. Morphol*, 258: 327–335.

In: Pesticides: Characteristics, Uses and Health Implications ISBN: 978-1-62081-577-9
Editors: A. Abrego Gomez and E. Lugo de Ortega

Chapter 2

ANALYTICAL METHODS FOR PESTICIDES DETERMINATION IN BIOLOGICAL MATRICES: A REVIEW

Laure Wiest* and Cécile Cren Olivé
Université de Lyon - Institut des Sciences Analytiques,
Département Service Central d'Analyse,
Equipe TRACES, CNRS, Université Lyon, Villeurbanne, France

ABSTRACT

Pesticides are among substances that have been identified as potential Endocrine Disrupting Chemicals (EDCs). The availability of validated exposure data is a critical component for assessing the causal relationships between exposure to EDCs and health effects. Although more and more data are available on these compounds, lots of questions about their ways of action are not answered yet and it is important to keep on studying them. The development of reliable multi-residue analytical methods is necessary to undertake this kind of study, especially on biological matrices such as plasma, serum, whole blood, urine, tissues, etc.

One of the main challenges in the development of analytical methods in biological matrices is the small amount of sample available. Therefore, special attention must be paid to the miniaturization of each step of the analytical method. Besides, as the number of samples in biomonitoring studies is very high, analytical methods developed have to be as fast as possible. Finally, studies on the behavior of pesticides in biological matrices not only require great sensitivity on a limited number of compounds but also information on untargeted molecules, such as their metabolites.

This chapter aims at giving an overview of current analytical methods of various families of pesticides in biological matrices, focusing on miniaturized sample preparation. The development of untargeted analytical methods and current data on pesticides transformation products will also be discussed.

---

* Corresponding author: Laure Wiest, Tel: (+33)478022277, Fax: (+33)478027187, E-mail: l.wiest@sca.cnrs.fr

1. Introduction

Pesticides are substances that can cause adverse effects by interfering in some way with the body's chemical messengers and/or hormones. Some belong to the list of potential endocrine disrupting chemicals (EDCs) with different families of chemicals: hormones, dioxins, polybromodiphenylethers, phtalates, parabens, alkylphenols, etc. For example, one of the most detected pesticides in the environment is atrazine that, among other effects, alters male reproductive tissues. The availability of validated exposure data is a critical component for assessing the causal relationships between exposure to EDCs and health effects [1]. Although more and more data are available on these compounds, lots of questions about their ways of action are not answered yet and it is important to keep on studying them. The development of reliable and sensitive multi-residue analytical methods is necessary to undertake this kind of study, especially on biological matrices such as plasma, serum, whole blood, urine, tissues, etc. In such complex matrices, sample preparation is crucial to obtain the required sensitivity and robustness.

Sample pre-treatment can also be considered as the most polluting step of the whole analytical process because it deals with the crude sample, in which the analytes may exist in 1000-fold smaller quantities than bulk constituents, so, often, the use of organic solvents is required to enrich the target compounds selectively and remove potentially interfering matter [2]. The development of techniques to reduce solvent consumption is not new: Wan et al. [3] reviewed in 1996 pathways and techniques to minimize organic solvent usage in pesticide residue analysis.

This review aims at giving an overview of the most advanced analytical techniques in pesticide analysis and various sample preparation of biological matrices, from the most conventional to solvent-free methods. Finally, the problem of matrix effect and the validation of the method will be discussed.

2. Analytical Techniques

The ability to estimate low-dose human exposure to commonly used pesticides often is requested in epidemiologic studies and the number of pesticides possibly present is very important. Therefore, fast and robust methods are necessary that can measure many analytes in the same sample. Analytical methods reviewed in the following section are the most sensitive and also the most universal to detect a large number of compounds.

2.1. Target Analysis

In order to achieve the sensitivity and the selectivity required to analyze traces of pesticides in very complex matrices such as biological matrices, chromatographic methods have to be used. Chromatographic methods coupled with mass spectrometry allow the best selectivity. Therefore, Gas Chromatography coupled with Mass Spectrometry (GC-MS) and high performance Liquid Chromatography coupled with tandem Mass Spectrometry (LC-MS/MS) are the most common analytical techniques used. The GC-MS technique is far older than LC-MS/MS. The first instrument was commercialized in the 1960s whereas the first LC-MS instrument, which is less selective than GC-MS, appeared in the 1990s and LC-MS/MS in 2000. As a result, GC-MS has been extensively used before 2000 to analyze pesticides in biological matrices.

The most studied groups of pesticides are organochlorine (OC), organophosphorous (OP) and pyrethrinoids (Pyr) also called "persistent pesticides". Due to their physico-chemical properties, the chromatographic technique of choice for these compounds is gas chromatography. The most used GC column is a low-polar fused silica coated with 5 % diphenyl/95% dimethyl polysiloxane stationary phase, usually called "-5" or "-5MS" from various suppliers [4-15]. The "-5MS" columns are specially made for low-bleed performance and are recommended for GC-MS. Classical dimensions of columns are 30 m length, 0,25 mm Internal Diameter (I.D.) and 0,25 μm film thickness, with helium as a carrier gas at constant flow of 1 mL/min. For multi-residue analysis, the total run time is usually of 30 min, leading to a final temperature at 300°C. The major way to decrease this run time that has been developed in the last 10 years is the use of narrow-bore columns. This is what is called Fast GC [16]. The I.D. of these columns is reduced, typically inferior to 0.2 mm, leading to an increase of the efficiency per length. Therefore, the length can be decreased in order to yield the same plate number in a shorter time, that is, inferior to 10 min. One drawback of these columns is a lower sample capacity which can affect the sensitivity of the method. Another important parameter in GC analysis is the injection volume, which is usually 1 or 2 μL in splitless mode. Nevertheless, in order to increase the sensitivity of the method, Cazorla et al. [8] used a programmed temperature vaporizing (PTV) injector operated in the large-volume injection mode, which allowed them to inject 10 μL. With this system, it is possible to use injection volume up to 40 μL [13]. This volume implies the use of a guard column before the analytical column and special attention has to be paid to matrix effect (see section 4). Finally, injection solvents are various, often depending on the sample preparation. Mastovska et al. [17] evaluated six common solvents, acetonitrile, acetone, ethyl acetate, hexane, toluene and isooctane, leading to the conclusion that acidified acetonitrile is the most suitable solvent for multiresidue pesticide analysis.

Regarding the detection, GC coupled with Electron Capture Detection (ECD) is the most common system used for halogenated pesticides, because of its lower prize compared to GC-MS [18]. But GC-ECD is less selective and sensitive than GC-MS and as single-quadrupole analysers are more and more affordable and easy to use, it is nowadays the technique most chosen. But these instruments show limitations in the analysis of traces in complex matrices: low resolution and low acquisition speed [19]. Low resolution implies low selectivity. To increase the resolution, ion trap quadrupoles were successfully used by Barr et al. [20], Perez et al [6] and Salihovic et al. [7]. Most of Limits of Quantification (LOQ) obtained were inferior to 0.1 ng/mL in plasma, which is a good performance. Another way to increase the selectivity in GC-MS, is the use of tandem mass spectrometry that can be performed using ion trap [8, 21] or triple quadrupole mass analyzers [15, 22]. LOQ obtained with these systems are also very low, but parent ions are often fragment ions which makes difficult to find suitable product ions, especially for molecules with mass inferior to 200 Da.

Even though GC-MS is the standard to analyze OC, OP and Pyr, their metabolites are often less volatile and more polar, characteristics that require LC-MS [23]. Due to lower chromatographic resolution and lower selectivity compared to GC-MS, LC-MS was rarely used for pesticide residue analysis [24, 25], but now that system of LC coupled with tandem mass spectrometry are affordable and more and more sensitive, LC-MS/MS is increasingly used. Adler et al. [26] selected 500 pesticides and compared the analytical performance of GC-MS and LC-MS/MS for these pesticides. According to this article, even if both techniques are needed, LC-MS/MS is suitable for a larger number of pesticides, is often more sensitive and more selective than GC-MS, which makes it the preferred technique, especially for multiresidue analysis.

Except for very polar compounds, most of LC analysis of pesticides are based on C_{18} reversed phase column from various suppliers [24, 27-29]. A better separation of very polar compounds can be obtained with HILIC (Hydrophilic Interaction Liquid Chromatography) columns whose coating is functionalized silica with hydrophilic groups. Odetokun et al. [30] developed an analytical method of dialkylphosphates in urine by HILIC-MS/MS, avoiding the time- consuming step of derivatization necessary for GC-MS analysis [5, 10, 31]. Typical dimensions of LC column are a length of 100 or 150 mm, an internal diameter of 2.1 or 4.6 mm and particles size of 3 µm. However, one of the main advances in LC in the last 5 years is the use of columns packed with sub-2 µm particles, which is called ultra-high-pressure liquid chromatography (UHPLC) [32] which allows better resolution and shorter analysis time. For example, Cazorla-Reyes et al. [8] developed an analytical method of more than 85 pesticides in urine by UHPLC-MS/MS. A C_{18} column with 1.8 µm particles was used, which allowed the separation of all the compounds in less than 10 min. A column packed with the same particle size was also used by Will et al. [33] to separate 11 pharmaceuticals and 14 pesticides in mussels. The major drawback of this technique is the backpressure which is higher than 400 bars, that is higher than the pressure that a conventional LC system can support. Another alternative is the use of fused core columns, which generate backpressure about two times lower than do columns packed with sub-2 µm particles with a comparable efficiency [34]. Wang et al. [35] developed an analytical method of 150 pesticides in grains with a core-shell particle column and compared it to conventional LC. This study showed that the use of a fused core column lead to a better sensitivity, a better repeatability and a shorter analytical run time than conventional LC. Finally, chromatographic resolution can be increased by reducing the I.D. of the column. This can be achieved with nano-liquid chromatography (nano-LC) system in which capillary columns of 100 µm I.D. are used. Nano-LC was successfully applied by Buonasera et al. [36] to the analysis of 8 OP in baby foods at low concentrations, with only an Ultra-Violet detection. Regarding the detection, atmospheric pressure ionization (API)-MS/MS is commonly used. Two types of API exist: electrospray ionization (ESI) and atmospheric pressure chemical ionization (APCI). As ESI is suitable for a larger number of compounds, it is becoming the reference [23]. APCI can be more efficient than ESI to ionize the most apolar pesticides. Panuwet et al. [37] successfully used it to analyze atrazine and its metabolites in urine at concentrations as low as 0.1-0.5 ng/mL. There are two modes of ionization in API: positive (addition of a proton) and negative (loss of a proton). Positive mode is suitable for a larger number of compounds than negative mode but it implies that there is more noise and possible interference in positive mode than negative mode. To ensure a good selectivity in LC-MS/MS, two Multiple Reaction Monitoring (MRM) transitions are needed for each compound and the ratio of the two transitions is also used as confirmation. For pesticides with poor fragmentation, confirmation by triple quadrupole instruments is sometimes not easy due to a low sensitive second MRM transition. Most recent tandem mass spectrometers give the possibility to acquire data dependent scan, also called information-dependent data acquisition. For example, Garcia-Reyes et al. [38]

used a 3200 QTrap from ABSciex which allowed combining a survey scan acquired in the MRM operation mode (triple quadrupole) with another survey scan in a Product Ion mode. Consequently, instead of a second MRM transition, a full spectrum of fragmentation is acquired. This new technology offers a gain of sensitivity but also the possibility to make libraries in LC-MS/MS which was a major drawback of this analytical technique.

2.2. Pesticides Metabolites and Untargeted Analysis

Pesticides are usually transformed through small changes from the parent pesticide molecule to complete mineralization to carbon dioxide, water, chloride, phosphate, etc. There are a large number of metabolites that have already been studied which were reviewed by Martinez Vidal et al. [39]. Time-of-flight MS (ToF-MS) screening has gained popularity in this field thanks to its capacity to acquire full-scan data with accurate mass. This acquisition allows not targeted screening which means that afterward, it is possible to check for the presence of a pollutant which has not been searched. Hernandez et al. [40] reviewed in 2008 the applications of LC-ToF and LC-Q-ToF to the determination of pesticides metabolites, illustrating the great potential of these techniques.

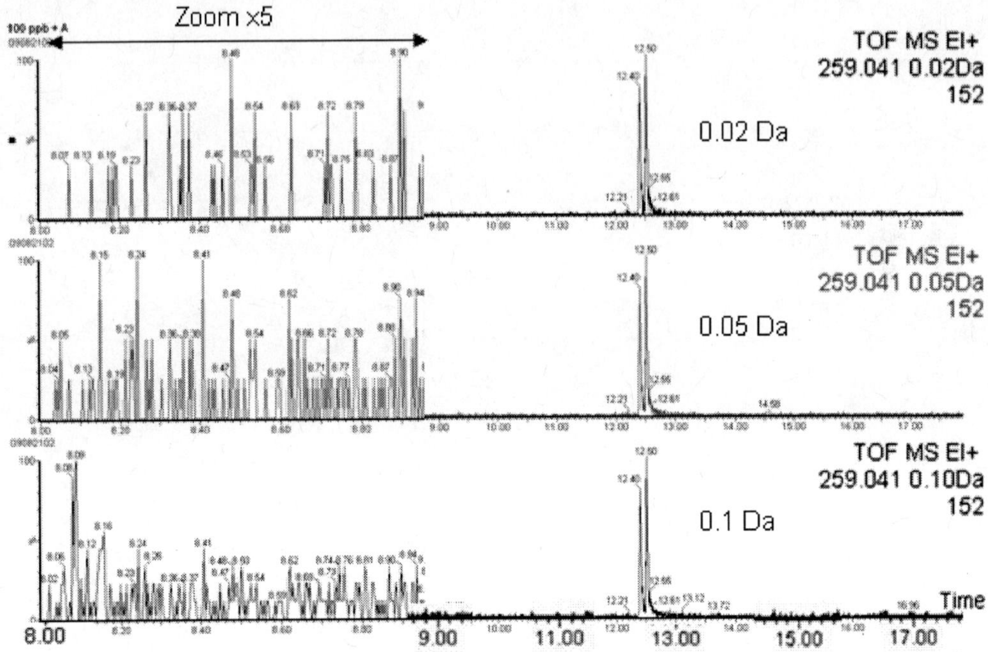

Figure 1. Influence of mass window width in GC-ToF analysis.

Furthermore, the dynamic range of the first LC-ToF or GC-ToF instruments was to narrow to allow quantification of pesticides at trace concentrations. This is not true anymore thanks to new technologies. Wiest et al. [41] successfully used a GC-ToF for the determination of 60 pesticides in honeybees, thanks to a new technology called Dynamic Range Enhancement. Moreover, a narrow mass window of 0.02 Da allowed a decrease of the noise by a factor of 3, leading to good limits of quantification. This point is illustrated in

Figure 1. Simplified instrument set-up, and reduced compromise in method performance when increasing the number of compounds contained within the scope of the method make LC-Q-ToF and GC-ToF more and more used analytical techniques.

3. Sample Preparation

It is well-known now that a conventional injection procedure, a single analytical column and a selective detector often does not suffice to recognize and quantify all analytes of interest in a sample [42]. The sample preparation is a crucial step of an analytical method. Furthermore, regarding certain matrices, small amount of sample is available and the environmental pressure demands procedures with less and less solvent used.

Table 1. Analytical methods in liquid matrices: Conventional sample preparation techniques

Analyte	Matrix	Vol (µL)	Extraction	Final vol (µL)	LOQ ng/mL	Analytical Technique	Ref
3 multiclass pesticides and their metabolites	Rat plasma and urine	500	SPE C_{18}	500	50-150	HPLC-UV	[54]
29 organophosphorous	Blood and serum	2000	SPE Oasis HLB	100	10-50	GC-MS	[52]
61 multiclass pesticides	Serum	1000	SPE Oasis HLB SPE Oasis MCX	100	10 5-100	GC-MS and LC-MS	[4]
25 multiclass pesticides (OC, OP, Pyr) and PCB	Blood	6000	LLE (hexane)	300	2-5	GC-MS	[45]
24 multiclass pesticides and metabolites	Serum and plasma	4000	SPE Oasis HLB	10	0.003-0.06	GC-HRMS (10000)	[20]
10 OP metabolites	Urine	2000	SPE Oasis HLB β-glucuronidase/ sulfatase	100	5-25	LC-MS/MS	[55]
4 OP metabolites and conjugates	Urine	500-1000	β-glucuronidase/ sulfatase 3-fold dilution		0.5 5-20	LC-2D-MS/MS LC-MS/MS (Ion Pair)	[57]
12 OP and Pyr metabolites 6 herbicides and DEET	Urine	2000	SPE Oasis HLB β-glucuronidase/ sulfatase	50	0.3-3	LC-MS/MS	[56]
16 sulfonyl urea	Urine	2000	SPE Oasis HLB	50	0.05-0.1	LC-MS/MS	[27]
6 OP metabolites	Urine	5000	LLE (diethylether-Ethyl Acetate)	80	2	LC-MS/MS	[28]
4 OP metabolites	Urine	5000	LLE (diethylether–ACN)SPE Florisil PSA	200	0.5-1	GC-MS	[5]

Analyte	Matrix	Vol (μL)	Extraction	Final vol (μL)	LOQ ng/mL	Analytical Technique	Ref
4 Pyr metabolites	Urine	5000 2000 5000	SPE Oasis HLB + Deriv LLE + Deriv SPE Oasis HLB	5000 250 5000	0.03-12	GC-MS GC-HRMS LC-MS/MS	[53]
10 organophosphorous	Serum	200	Precipitation with acetonitrile	200	0.25-1.25	LC-MS	[24]
Atrazine and its metabolites	Urine	200	SPE online (Strata X)	500	0.5-5	LC-MS/MS (APCI)	[65]
4 organophosphorous	Blood	100	Precipitation with MeOH:Zn-sulfate	100	0.5-1250	LC-MS/MS	[29]
320 multiclass pesticides and metabolites	Blood	1000	SPE Oasis WAX	80	10-10000	LC-IT-MS	[64]
Atrazine and its metabolites	Urine	1000	SPE Strata X-C	100	0.1-0.5	LC-MS/MS (APCI)	[37]
17 multiclass pesticides	Plasma	2000	SPE ABS ELUT-Nexus	10	0.02-0.2	GC-HRMS (10000)	[6]
6 OP metabolites	Urine	600	SPE Strata X-AW 96-well plate	50	0.04-1.5	hilic-LC-MS/MS	[30]
5 organophosphorous	Serum albumin and blood	500	Enzyme-Digestion LLE (hexane)	75	nd	GC-MS	[44]
5 OC pesticides and other pollutants	Plasma	500	SPE Oasis HLB Clean-up Silica gel	25	0.01-1	HRGC-HRMS (10000)	[7]
More than 200 multiclass pesticides and their metabolites	Urine	5000	SPE C18	2000	0.003-1.45 0.011-3.49	GC-MS/MS UHPLC-MS/MS	[8]
Atrazine, Vinclozolin M2, HPTE	Rat serum	1000	SPE Oasis HLB	50	0.03-0.6	UHPLC-MS/MS	[63]

OC=organochlorine; OP=organophosphorous; Pyr=pyrethroid; HPTE=2,2-bis(p-hydroxyphenyl)-1,1,1-trichloroethane ; SPE=Solid Phase Extraction ; LLE=Liquid Liquid Extraction.

Table 2. Analytical methods in liquid matrices: Micro-extraction techniques

Analyte	Matrix	Vol (μL)	Extraction	LOQ ng/mL	Analytical Techniques	Ref
4 OP	Blood, plasma, cerebrospinal fluid	300	HeadSpace SPME Polyacrylate fiber	6-100	GC-NPD	[47]
Parathion and metabolites	Blood and urine	100	direct immersion SPME Carbowax/ divinylbenzene	10-50	GC-MS	[50]

Table 2. (Continued)

Analyte	Matrix	Vol (µL)	Extraction	LOQ ng/mL	Analytical Techniques	Ref
5 OC and 18 PCB	Serum	500	HeadSpace SPME PDMS fiber	0.003-0.05	GC-ECD	[49]
4 dinitro phenol pesticides	Plasma	1000	Hollow-Fibre LPME	50-100	LC-MS	[25]
7 phenylurea and triazine herbicides	Milk	4000	Microwave assisted ionic liquid DLLME	1-4	HPLC-UV	[46]
6 triazoles	Milk	5000	DLLME	4-58 13-180	GC-FID GC-MS	[48]
6 Pyr metabolites	Urine	1000	Hollow-Fibre LPME	4-50	GC-ECD	[31]
16 OC, 7 Pyr and other pollutants	Porcine plasma	4000	Hollow-Fibre LPME	0.05-0.4	GC-MS/MS	[21]

OC=organochlorine; OP=organophosphorous; Pyr=pyrethroid; PCB=Polychlorinated biphenyls ; SPME=Solid Phase MicroExtraction; PDMS=Polydimethylsiloxane; LPME= Liquid Phase MicroExtraction; DLLME=Dispersive liquid liquid microextraction.

Consequently, there is an increase of the development of solventless techniques such as microextractions [2, 43].

The objective of this section is to review robust classical extraction techniques (Table 1) but also promising microextractions (Table 2) in liquid and solid matrices (Table 3).

Table 3. Analytical methods in solid matrices

Analyte	Matrix	Mass (mg)	Extraction	Final vol (µL)	LOQ ng/g	Analytical Techniques	Ref
25 multiclass pesticides (OC, OP, Pyr) and PCB	Dog hair	200	SLE with hexane and hexane/ethyl acetate	300		GC-MS	[45]
11 multiclass pesticides and their metabolites	Hair	50	SLE with hexane	100	31-5880	GC-MS	[9]
4 OP metabolites	Hair	100	Ultrasonic SLE with methanol	50	5-20	GC-MS	[10]
6 multiclass pesticides	Hair	50	SLE methanol/ phosphate Hollow-Fiber SPME		0.3-15	HPLC-UV	[90]
22 multiclass pesticides	Hair	50	SLE with ACN DI-SPME PDMS/DVB		0.05-10	GC-MS/MS	[22]
21 OC, 22 PCB and 12 PBDE	Organ tissue	5000	PLE(DCM/acetone) Clean-up GPC and SPE	200	1-5	GC-MS	[75]

Analyte	Matrix	Mass (mg)	Extraction	Final vol (µL)	LOQ ng/g	Analytical Techniques	Ref
21 OC, 7 PCB and 2 PBDE	Breast tissue	500	SLE with hexane Clean-up with LC	1000	5-25	GC-MS/MS	[15]
11 multiclass pesticides and their metabolites	Meco nium	500	SLE with methanol/ Phosphate SPE C18	100	400	GC-MS	[51]
8 OC, 6 PCB	Breast tissue	500	SLE with hexane Clean-up with LC or SPE	500	10-250	GC-ToF	[79]
Paraquat	Brain tissue	15	Microwave assisted solvent extraction	150	7	HPLC-MS	[73]
8 Pyr	Benthic Inver tebrates	500	Matrix Solid Phase Dispersion	100	1.5-12	GC-ECD	[11]
2 Pyr	Fish tissue	5000	LLE with ACN and salts Dispersive SPE PSA/C18	500	1	GC-MS	[12]
14 multiclass pesticides and other pollutants	Mussels	1000	PLE (ACN/methanol) SPE Isolute ENV+	200	1-10	UHPLC-MS/MS LC-ToF	[33]
13 multiclass pesticides	Fish tissue	3000	LLE with ACN, hexane and salts	3000	1-7	UHPLC-MS/MS	[70]
22 OC and 7 PCB	Fish tissue	5000	LLE with ACN, THF Dispersive SPE PSA	10000	1-10	GC-MS	[14]
Atrazine, Vinclozolin M2, HPTE	Rat testis	1500	LLE with ACN, hexane and salts	105	0.15-20	UHPLC-MS/MS	[71]

OC=organochlorine; OP=organophosphorous; Pyr=pyrethroid; PCB=Polychlorinated biphenyls ; HPTE=2,2-bis(p-hydroxyphenyl)-1,1,1-trichloroethane ; ACN= acetonitrile ; THF=tetrahydrofurane ; PSA=Primary Secondary Amine ; SLE=Solid liquid extraction ; SPE=Solid Phase Extraction ; LLE=Liquid Liquid Extraction ; SPME=Solid Phase MicroExtraction; GPC= gel permeation chromatography ; PDMS/ DVB=Polydimethylsiloxane/ divinylbenzene; LPME= Liquid Phase MicroExtraction; DLLME=Dispersive liquid liquid microextraction.

3.1. Liquid Matrices

3.1.1. Sample Pre-Treatment

Regarding blood, plasma, serum and milk, protein precipitation is usually the first step of the sample preparation. However, it is sometimes used as a one step sample preparation. Inoue et al. [24] and Salm et al. [29] successfully used ACN and zinc sulphate: methanol mixture respectively for protein precipitation of serum and blood and extraction of OP. These two simple methods gave good recoveries and low LOQ with a total extraction time inferior to 10 min and small solvent consumption (100-200 μL). The importance of this step was demonstrated by Yue et al. [44]. Three extraction methods were compared: a (a) direct liquid-liquid extraction with hexane, (b) the same LLE after precipitation with acetone and (c) the same LLE after reduction, alkylation and typical digestion for the extraction of 5 OP in blood. Recoveries obtained with the procedure (a) were from 20 to 45%, (b) were from 20 to 72% and (c) were from 43 to 103%. Protein precipitation is usually performed with sodium chloride, alone [45, 46] or in combination with a solvent [47, 48] but also with a phosphate buffer [49-51], saturated ammonium sulfate [20] or acetonitrile [52].

Concerning urine analysis, the presence of possible glucuronide or sulfate-conjugated metabolites should be avoided. Consequently, a hydrolysis step is usually performed with hydrochloric acid [5, 28, 31, 53], acetic acid [54] or formic acid [30]. In order to get a more selective hydrolysis, an enzymatic hydrolysis instead of a simple hydrolysis can be performed with the use of β-glucuronidase and/or sulfatase enzymes [55-57].

3.1.2. Liquid Liquid Extraction Techniques

3.1.2.1. Liquid-Liquid Extraction

Liquid-liquid extractions (LLE), with Solid Phase Extractions (SPE) are the most classical preparation techniques. Regarding apolar compounds, good recoveries can be obtained with LLE. For example, hexane was used to recover apolar pesticides from blood [44, 45]. Concerning polar molecules, good extraction efficiency is often difficult to obtain with one solvent. Methods are consequently developed using mixtures of solvent, leading to more matrix interferences in the extract and the addition of a clean-up step. For example, mixtures of diethylether-ethyl acetate [28] and diethylether–acetonitrile [5] were used to extract OP metabolites from urine. Finally LLE requires often high amount of solvent and is time–consuming. As a result, liquid phase microextractions are more and more developed.

3.1.2.2. Liquid-Phase Microextraction

Liquid-phase microextraction (LPME) techniques can be classified into three categories: single-drop microextraction (SDME), hollow-fiber LPME (HF-LPME) and dispersive liquid–liquid microextraction (DLLME) [58]. Although these techniques are mainly used for the extraction of inorganic and organic compounds in water, applications of them to biological samples are more and more published.

In SDME, analytes are extracted from a liquid sample in a microdrop of solvent (usually few microliters) which is formed at the tip of a microsyringe needle. After extraction, the microdrop is retracted back into the microsyringe and injected into GC or LC. All the possible modes of SDME were recently reviewed by Jain et al. [59]. One of the disadvantages of SDME is that the drop can be unstable.

To increase this stability, hollow-fiber membrane (HF) LPME was developed [60]. With this method, the extraction solvent is immobilized in the wall pores of the HF. Lin et al. [31] developed this technique for the extraction of Pyr metabolites in urine and compared it to conventional LLE and SPE. Lower LOD were achieved with HF-LPME, mostly thanks to less matrix interferences in the volume injected.

A multiresidue analysis of 40 organic pollutants with various physico-chemical properties in porcine plasma by HF-LPME-GC-MS/MS was developed by Manso et al. [21]. After optimization with an experimental design, very good recoveries were obtained for all compounds with LOQ lower than 0.5 ng/mL.

Another interesting, fast and easy technique is DLLME. In DLLME, analytes are extracted into fine droplets (dispersion) of the extracting solvent made in water. To achieve dispersion, a second solvent (the dispersing solvent) is used [61]. Like HF-LPME, this technique allows to obtain high enrichment factor (EF) leading to low LOQ. Farajzadeh et al. [48] used this method for the extraction of triazole in milk. First a salting-out liquid liquid extraction of 5 mL of milk was done with acetonitrile and then the resulting phase of acetonitrile was extracted with chloroform. Good method performance was obtained with enrichment factors in the range of 156-180. One of the crucial step of the DLLME is a good separation of the extraction solvent. As ionic liquids have very special properties for extraction, their use in various liquid extraction has increased for few years [62]. Gao et al. [46] developed an extraction of phenylurea and triazine herbicides in milk with imidazolium ionic liquids. Ultrasonication was used to disperse the ionic liquid into the sample and the extraction was assisted by microwave. This procedure took only 7 min per sample, giving recoveries higher than 88% and interesting LOQ with only UV detection. Although LPME techniques are quite recent, lots of applications have already been published. Indeed all these techniques are simple, robust, quick, and cost-effective and require low amount of solvent.

3.1.3 Solid Phase Extraction Techniques

3.1.3.1. Solid Phase Extraction

One of the advantages of Solid Phase Extraction (SPE) is the concentration of the sample leading to low method LOQ, inferior to 0.1 ng/mL. Table. 1 shows the volume of sample and the volume of the final extract: concentration factors are between 10 [4] and 400 [20]. The most universal and used phase for SPE is Oasis HLB from Waters. Oasis HLB is a spherical Hydrophilic-Lipophilic-Balanced water-wettable reversed-phase made from two monomers, hydrophilic N-vinylpyrrolidone and lipophilic divinylbenzene. This sorbent shows retention of both polar and hydrophobic compounds, and consequently is one of the most suitable for multi-residue analysis. Table 1 shows that this sorbent was used to extract very different compounds in blood, serum, plasma and urine: OP [52], OP metabolites [55], sulfonyl urea pesticides [27], Pyr metabolites [53], OC [7] and multiclass pesticides [4, 20, 56, 63]. However, Cazorla et al. [8] compared Oasis HLB and C_{18} cartridges for the extraction of 200 pesticides in urine. Performances were similar regarding LC-amenable pesticides but concerning GC-amenable pesticides which are more apolar, better extraction efficiency was obtained with C_{18}. Pérez et al. [6] also used a more apolar phase for the extraction of 17 multiclass GC-amenable pesticides in plasma. Finally, Dulaurent et al. [64] studied mixed-mode anion exchange (Oasis WAX, MAX) and more hydrophobic sorbents (Oasis HLB, Varian Plexa) for the extraction in blood of a list of 320 pesticides, most of them having neutral or acidic properties: Oasis WAX was more powerful than MAX and more efficient than Oasis HLB and Varian Plexa regarding acidic compounds such as dialkylphosphates.

Consequently, compared to LLE, SPE is more universal and requires less solvent. SPE is a quite long extraction (between 30 min and 1 hour per sample) but automation is possible, which makes it less time-consuming. To increase their throughput using SPE, Odetokun et al. [30] used a 96 well plate with weak anion exchange cartridges for the extraction of dialkylphosphates in 600 μL of urine, making possible the extraction and analysis of 1152 samples per week. Another way to completely avoid time-consuming sample preparation is the development of online SPE coupled with HPLC. An online SPE-HPLC-MS/MS method was developed by Panuwet et al. [65] for the simultaneous extraction and analysis of atrazine and 6 metabolites in urine. Good recoveries and accuracy were obtained. LOQs were in the ng/mL range with only 200 µL of urine. Another way to use small volume of sample and facilitate automation is miniaturization: Micro-extraction by packed sorbent and Solid Phase Micro Extraction are promising techniques.

3.1.3.2. Micro-Extraction by Packed Sorbent

Recently, a miniature SPE was developed, which is called Micro-extraction by packed sorbent (MEPS). Principles and steps of MEPS are the same as SPE: 1 to 5 mg of sorbent are inserted into a syringe, instead of about 100 mg into a cartridge and 20 to 50 µL of solvent are used for elution instead of 2 to 5 mL, making MEPS a solventless technique. MEPS is also a cost-effective method as MEPS cartridges can be reused for about 100 injections, depending on the matrix. A complete review of recent advances in MEPS for bioanalysis was published [66]. It reports also the use of MEPS online with LC and GC. MEPS syringes are commercially available from SGE Analytical Science but only in a restricted number of phases. Indeed, Tournier et al. [63] used C_{18} MEPS for the extraction of bisphenol A, pesticides, hormones and sulfate and glucuronide estrogens in serum. Very good recoveries were obtained except for sulfate and glucuronide conjugates which are very polar compounds. A more hydrophilic phase would probably have given better results. Besides, a significant decrease of matrix effect was observed compared to conventional SPE.

3.1.3.3. Solid Phase Micro Extraction

Solid Phase Micro Extraction (SPME) has been widely used in bioanalytical methods for the extraction of solvent and drugs [67] but not that much regarding pesticides [68, 69]. In SPME, analytes are sorbed onto the surface of a coated silica fiber and then desorbed for injection into GC or LC. Two types of extraction exist: direct immersion SPME (DI-SPME) and head space SPME (HS-SPME). In DI-SPME, the coated fiber is directly inserted into the liquid sample and in HS-SPME, the fiber is not in contact with the sample, it is placed above a vial of the sample which is heated to volatize the analytes. When SPME is coupled with LC, compounds are desorbed in a suitable solvent whereas when coupled with GC, compounds are directly desorbed from the fiber into the GC injector. Therefore, this technique is more used for the analysis of volatile and semi-volatile compounds with GC system. To the best of our knowledge, concerning pesticides in liquid biological matrices, SPME is more often developed on a single compound except in the following articles: Gallardo et al. [50] used DI-SPME for parathion and its metabolites in blood and urine, and HS-SPME was used by Tsoukali et al. [47] and Lopez et al. [49] for the extraction of OP in various biological matrices and OC in serum respectively. Experimental conditions reported in the literature for SPME are very different. Consequently, different parameters including fiber type, temperature, salt addition, sample pH, extraction and desorption time have to be optimized. An experimental design is recommended for this optimization [49]. Besides, low LOQ were obtained with only small volumes of samples (lower than 500 µL). SPME was also used for the extraction of hair (see Table 3). To conclude, SPME is a very interesting method as it is solvent-free when coupled with GC but it is difficult to use for multiclass pesticides analysis, due to compound dependent experimental conditions.

3.2. Solid Matrices

3.2.1. Sample Pre-Treatment

Before extraction, comminution of solid matrices is very important in order to obtain very thin particles of matrix, almost like a powder that can be easily extracted. Concerning hair sample, pulverization in a ball mill homogenizer is usually performed [9, 10, 22]. Regarding tissues, samples are freeze-dried [33] or mechanically homogenized with a high speed disperser like Polytron [14] or ultra-turrax [70, 71]. Another system of homogenization, called Geno/Grinder, was recently applied to pesticide analysis [72]. The matrix and solvent are first introduced into centrifuge tube with grinding balls, and the system shake the tube vertically at high speed. This system is interesting because it allows simultaneous comminution and extraction of the sample and simultaneous extraction of at least nine samples.

3.2.2. Solvent Extraction

Conventional Solid Liquid Extraction with one organic solvent is less and less used because of a too long preparation time. For example, Posecion et al. developed an extraction of 11 multiclass pesticides and their metabolites in human hair with hexane, which lasted 6 hours [9]. Also regarding human hair matrix, the extraction time of the method published by Salquebre et al. for the analysis of 22 multiclass pesticides was of 12 hours [22]. The use of ultrasonication followed by mechanical shaking allowed Tsatsakis et al. [10] to reduce the extraction time to 4 hours and a half.

Consequently, to decrease the preparation time, assisted extractions with microwave or pressure were developed. Table 3 shows some applications of these extractions to tissues. Microwave assisted solvent extraction (MAE or MASE) was used to extract paraquat from brain tissues with acetic acid [73]. By applying microwave energy to the extraction solvent, the highly localized temperature and pressure cause heating of the matrix and migration of target analytes from the sample material to the surrounding solvent rapidly [74]. Pressurized Liquid Extraction (PLE) is another fast extraction technique which uses high-temperature and pressure. Depending on the polarity of the solvents used, it can be applied to a large number of compounds, from the most apolar [75] to the most polar [33]. PLE is a robust method of extraction, more efficient than conventional soxhlet but one of its disadvantages is that a clean-up is required before injection into a chromatographic system, leading to a multi-step sample preparation.

Generally speaking, regarding the different organic solvent used for extraction, a comparison of three solvents, acetonitrile, ethyl acetate and acetone was published by Mastovska et al. [17] leading to the conclusion that acetonitrile is the most suitable for sample preparation in pesticide multiresidue analysis. Indeed, acetonitrile is the solvent used in one of the most universal extraction method to analyse a wide range of pesticides: the "QuEChERS method".

This method consists in two steps, solid-liquid-liquid extraction, and purification by dispersive Solid Phase Extraction (dSPE) (see section 3.2.3). QuEChERS stands for Quick, Easy, Cheap, Rugged and Safe. This method was first introduced by Anastassiades et al. in 2003 [76]. The original method was developed in order to extract pesticides from fruits and vegetables. Since then, it has been applied to a large range of matrices [77] including tissues and blood [78]. Initially, the first step consisted in salting-out liquid-liquid extraction, using acetonitrile, magnesium sulfate to remove water and NaCl salts. Then acetate salts or citrate salts were added in order to buffer the liquid-liquid extraction and avoid the degradation of base-sensitive pesticides. This last method was successfully applied by Rawn et al. [12] to the extraction of pyrethrins and Pyr in fish tissues.

To remove the most apolar interferences, a small amount of a second apolar solvent can be added: tetrahydrofurane was used by Norli et al. [14] to extract OC pesticides from fish tissues and hexane was used by Lazartigues et al. [70], Wiest et al. [41] and Pouech et al. [71] to extract pesticides of various physico-chemical properties from fish tissue, honeybees and rat testis respectively. QuEChERS has proved to be a very efficient and robust method. Another advantage is that as the solvent extraction is acetonitrile, the final extract can be injected either in GC or LC [41].

3.2.3. Clean-Up And/Or Pre-Concentration

After solvent extraction, a clean-up step is often required. Nevertheless, it is not always useful and it can sometimes decrease the recovery of extraction: Lazartigues et al. [70] and Pouech et al. [71] used only a QuEChERS based extraction with hexane without any further purification step. Very various clean-up techniques do exist.

All the techniques presented in section 3.1 can be applied to the solvent extract like SPE [33, 75, 79]. One of the main interferences to remove when analyzing tissues is lipids. Chromatographic separations can be an efficient tool for purification: gel permeation chromatography (GPC) was used by Saito et al. [75] to purify the extract from PLE and LC was used by Hernandez et al. [15] to purify an extract in hexane.

Another interesting sample preparation technique which allows homogenization and clean-up at the same time is matrix solid phase separation (MSPD). The principle of this technique is the same as solid phase extraction.

The sample is directly in contact with a phase adsorbent in a mortar and the mixture is then eluted: interferences are retained on the adsorbent and compounds of interest are in the elution solvent. MSPD was used by Ding et al. [11] to determine cypermethrin residues in carp tissues. But one of the difficulties of MSPD is to obtain a complete homogenous mixture with the matrix and the solid phase adsorbent.

Dispersive SPE (d-SPE) is a clean-up based on the same principle as MSPD: interferences are retained on a solid phase adsorbent. But it is applied to a liquid extract, which increases the contact between the adsorbent and interferences of the matrix.

Most used adsorbents are Primary and Secondary Amine bonded silica alone [14] or in combination with octadecyl bonded silica [12]. As d-SPE is the second step of the QuEChERS method, its applications to pesticide residue analysis are very numerous.

Finally, Tsai et al.[80] published an interesting comparison of d-SPE and DLLME as clean-up steps for quinolones extraction from swine muscles. Both give similar recoveries but DLLME allowed a lower amount of interferences in the final extract, leading to lower matrix effect.

4. MATRIX EFFECTS / QUANTIFICATION

GC-MS and LC-MS/MS are the most used techniques for multi-residue analysis of pesticides but they are both subject to strong matrix effects (ME) which can significantly reduce or enhance the analyte response and lead to wrong quantification. Consequently, solutions have to be found to reduce them and to take them into account for quantification. Most of them have been reviewed in 2003 by Ms Hajslová [81].

Regarding ME in pesticide analysis by GC, an innovative solution was published by Anastassiades et al [82] which consist in adding compounds called "analyte protectants" that interact more than pesticides with active sites present in the chromatographic system. There is no equivalent considering LC analysis but, according to Gros et al. [83], the dilution of sample extracts is an efficient strategy to reduce ion suppression.

Even if these solutions can reduce ME, it is not enough to apply external calibration in solvent to quantify compounds in complex matrices such as biological matrices without any error. Therefore three calibration strategies are possible: internal standard, standard addition and matrix-matched calibrations. Internal standard calibration is the most efficient method to correct ME but it requires the use of isotope labeled standards of the target analytes [84] that can be very expensive and sometimes unavailable. Standard addition is also an efficient method but very time-consuming especially when a large number of samples have to be analyzed and it requires a high amount of matrix.

Finally matrix-matched calibration is an efficient method to correct ME, only if a blank matrix is available and if the ion suppression/enhancement effects due to the matrix used for calibration are the same as the effects of the samples analyzed. Indeed, this last condition does not concerned blood or plasma but urine samples can be very different between each other due to dilution [18].Whatever the strategy chosen, these considerations prove that the accuracy of the method has to be studied and that the LOQ of a method has to be calculated in the matrix which is part of the validation of the method.

5. VALIDATION OF THE METHOD

The purpose of method validation is to ensure that an analytical methodology is accurate, specific, reproducible and robust over the specified range that a compound will be analyzed [85]. Numerous guidelines have been published regarding method validation [86] but some validation parameters are common. Limits of detection and/or limits of quantification, accuracy and recovery, repeatability and intermediate precision should be at least determined. One of the difficulties to perform a method validation is the high number of experiments required. It is therefore useful to define a validation plan that allows determining different parameters in combined experiments. Based on the ICH guideline [87], Wiest et al. [41] published a validation plan in 3 days, with 11 experiments per day (Figure 2), which allows the determination of linearity, recovery, repeatability and intermediate precision.

Another approach, based on accuracy profiles, developed by Feinberg Max [88] offers the advantage among others to simultaneously evaluate precision and trueness. Most used criteria of validation are recoveries between 60 and 120%, Relative Standard Deviation (RSD) lower than 20% for repeatability and lower than 25% for intermediate precision. But depending on the matrix and the concentration of the analyte, these criteria can be difficult to reach.

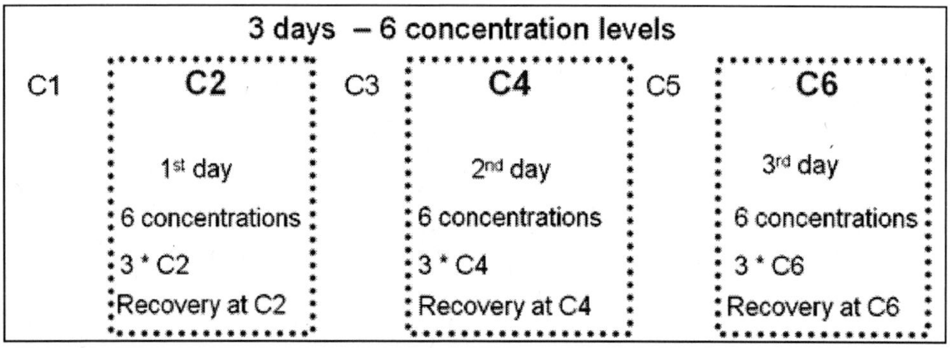

Figure 2. Validation strategy on 3 days.

The AOAC manual for the Peer-Verified Methods program [89] includes a table with estimated recovery data as a function of analyte concentration. For example, acceptable recoveries and RSD vary from 80-110% and 11% respectively, at 1 ppm to 40-120% and 30% at 1 ppb. Finally, even if LOD and LOQ of an analytical method are very important, there is no standardization of their calculations. Different approaches can be used: the visual inspection, the standard deviation of the response and the slope of the calibration curve and the signal-to-noise ratio convention (usually 3:1 for LOD and 10:1 for LOQ), among others. That's why any values of LOD and specially LOQ must be checked by replicates using samples spiked at concentrations in the range of these limits. In conclusion, Rambla-Alegre et al.[86] recently published a very interesting and complete review about validation of analytical method which title is "Is it really necessary to validate an analytical method?" and the answer is of course Yes.

CONCLUSION

This review gave an overview of the most advanced analytical techniques in pesticide analysis in biological matrices. The objectives of analysts have always been to develop reliable and sensitive analytical methods.

Now two more challenges are added: to develop solvent-less sample preparation and analytical methods that allow the determination of the largest number of compounds possible. Miniaturized analytical techniques (micro-extractions, nano-LC) and Time-of-Flight mass spectrometry are sure parts of the answers to these challenges.

REFERENCES

[1] World Health Organization, 2002, *Global assessment of the state-of-the-science of endocrine disruptors*, WHO/PCS/EDC/02.2, available at http://www.who.int/ipcs/publications/new_issues/endocrine_disruptors/en/.

[2] Farré, M., Pérez, S., Gonçalves, C., Alpendurada, M. F. and Barcelo, D. (2010). *TrAC Trends in Analytical Chemistry*, 29, 1347-1362.

[3] Wan, H. B. and Wong, M. K. (1996). *Journal of Chromatography A*, 754, 43-47.

[4] Lacassie, E., Marquet, P., Gaulier, J.-M., Dreyfuss, M.-F. and Lachâtre, G. (2001). *Forensic Science International*, 121, 116-125.

[5] Ueyama, J., Saito, I., Kamijima, M., Nakajima, T., Gotoh, M., Suzuki, T., Shibata, E., Kondo, T., Takagi, K., Miyamoto, K.-i., Takamatsu, J., Hasegawa, T. and Takagi, K. (2006). *Journal of Chromatography B*, 832, 58-66.

[6] Pérez, J. J., Williams, M. K., Weerasekera, G., Smith, K., Whyatt, R. M., Needham, L. L. and Barr, D. B. (2010). *Journal of Chromatography B*, 878, 2554-2562.

[7] Salihovic, S., Mattioli, L., Lindstrom, G., Lind, L., Monica Lind, P. and Bavel, B. v. (2011). *Chemosphere.*

[8] Cazorla-Reyes, R., Fernandez-Moreno, J. L., Romero-Gonzalez, R., Frenich, A. G. and Vidal, J. L. M. (2011). *Talanta*, 85, 183-196.

[9] Posecion, N., Ostrea, E., Bielawski, D., Corrion, M., Seagraves, J. and Jin, Y. (2006). *Chromatographia*, 64, 681-687.

[10] Tsatsakis, A. M., Barbounis, M. G., Kavalakis, M., Kokkinakis, M., Terzi, I. and Tzatzarakis, M. N. (2010). *Journal of Chromatography B*, 878, 1246-1252.

[11] Ding, Y., You, J. and Lydy, M. J. (2009). *Bulletin of Environmental Contamination and Toxicology*, 83, 388-392.

[12] Rawn, D., Judge, J. and Roscoe, V. (2010). *Analytical and Bioanalytical Chemistry*, 397, 2525-2531.

[13] Castillo, M., Gonzalez, C. and Miralles, A. (2011). *Analytical and Bioanalytical Chemistry*, 400, 1315-1328.

[14] Norli, H. R., Christiansen, A. and Deribe, E. (2011). *Journal of Chromatography A*, 1218, 7234-7241.

[15] Hernández, F., Portoles, T., Pitarch, E., Lopez, F. J., Beltran, J. and Vazquez, C. (2005). *Analytical Chemistry*, 77, 7662-7672.

[16] Domotorova, M. and Matisova, E. (2008). *Journal of Chromatography A*, 1207, 1-16.

[17] Mastovska, K. and Lehotay, S. J. (2004). *Journal of Chromatography A*, 1040, 259-272.

[18] Barr, D. B. and Needham, L. L. (2002). *Journal of Chromatography B*, 778, 5-29.

[19] Hernández, F., Portoles, T., Pitarch, E. and Lopez, F. J. (2011). *TrAC Trends in Analytical Chemistry*, 30, 388-400.

[20] Barr, D. B., Barr, J. R., Maggio, V. L., Whitehead Jr, R. D., Sadowski, M. A., Whyatt, R. M. and Needham, L. L. (2002). *Journal of Chromatography B*, 778, 99-111.

[21] Manso, J., Garcia-Barrera, T. and Gomez-Ariza, J. L. (2011). *Journal of Chromatography A*, 1218, 7923-7935.

[22] Salquèbre, G., Schummer, C., Millet, M., Briand, O. and Appenzeller, B. M. R. (2012). *Analytica Chimica Acta*, 710, 65-74.

[23] Hernández, F., Sancho, J. V. and Pozo, O. J. (2005). *Analytical and Bioanalytical Chemistry*, 382, 934-946.
[24] Inoue, S., Saito, T., Mase, H., Suzuki, Y., Takazawa, K., Yamamoto, I. and Inokuchi, S. (2007). *Journal of Pharmaceutical and Biomedical Analysis*, 44, 258-264.
[25] Hansson, H., Lagerstrom, M., Aberg, M. and Nilsson, U. (2009). *Talanta*, 79, 633-638.
[26] Alder, L., Greulich, K., Kempe, G. and Vieth, B. (2006). *Mass Spectrometry Reviews*, 25, 838-865.
[27] Baker, S., Olsson, A., Needham, L. and Barr, D. (2005). *Analytical and Bioanalytical Chemistry*, 383, 963-976.
[28] Dulaurent, S., Saint-Marcoux, F., Marquet, P. and Lachâtre, G. (2006). *Journal of Chromatography B*, 831, 223-229.
[29] Salm, P., Taylor, P. J., Roberts, D. and de Silva, J. (2009). *Journal of Chromatography B*, 877, 568-574.
[30] Odetokun, M. S., Montesano, M. A., Weerasekera, G., Whitehead Jr, R. D., Needham, L. L. and Barr, D. B. (2010). *Journal of Chromatography B*, 878, 2567-2574.
[31] Lin, C. H., Yan, C. T., Kumar, P. V., Li, H. P. and Jen, J. F. (2011). *Analytical and Bioanalytical Chemistry*, 401, 927-937.
[32] Nováková, L. and Vlcková, H. (2009). *Analytica Chimica Acta*, 656, 8-35.
[33] Wille, K., Kiebooms, J., Claessens, M., Rappé, K., Vanden Bussche, J., Noppe, H., Van Praet, N., De Wulf, E., Van Caeter, P., Janssen, C., De Brabander, H. and Vanhaecke, L. (2011). *Analytical and Bioanalytical Chemistry*, 400, 1459-1472.
[34] Guillarme, D., Ruta, J., Rudaz, S. and Veuthey, J.-L. (2010). *Analytical and Bioanalytical Chemistry*, 397, 1069-1082.
[35] Wang, J., Chow, W. and Cheung, W. (2011). *Journal of Agricultural and Food Chemistry*, 59, 8589-8608.
[36] Buonasera, K., D'Orazio, G., Fanali, S., Dugo, P. and Mondello, L. (2009). *Journal of Chromatography A*, 1216, 3970-3976.
[37] Panuwet, P., Restrepo, P. A., Magsumbol, M., Jung, K. Y., Montesano, M. A., Needham, L. L. and Barr, D. B. (2010). *Journal of Chromatography B*, 878, 957-962.
[38] Garcia-Reyes, J. F., Hernando, M. D., Ferrer, C., Molina-Diaz, A. and Fernandez-Alba, A. R. (2007). *Analytical Chemistry*, 79, 7308-7323.
[39] Martinez Vidal, J. L., Plaza-Bolanos, P., Romero-Gonzalez, R. and Garrido Frenich, A. (2009). *Journal of Chromatography A*, 1216, 6767-6788.
[40] Hernández, F., Sancho, J. V., Ibanez, M. and Grimalt, S. (2008). *TrAC Trends in Analytical Chemistry*, 27, 862-872.
[41] Wiest, L., Buleté, A., Giroud, B., Fratta, C. d., Amic, S., Lambert, O., Pouliquen, H. and Arnaudguilhem, C. (2011). *Journal of Chromatography A*, 1218, 5743-5756.
[42] Geerdink, R. B., Niessen, W. M. A. and Brinkman, U. A. T. (2002). *Journal of Chromatography A*, 970, 65-93.
[43] Nerín, C., Salafranca, J., Aznar, M. and Batlle, R. (2009). *Analytical and Bioanalytical Chemistry*, 393, 809-833.
[44] Yue, Y., Li, L., Liang, Y., Dong, L., Dong, Q. and Zhong, H. (2010). *Journal of Chromatography B*, 878, 3089-3094.
[45] Liu, S. and Pleil, J. D. (2002). *Journal of Chromatography B*, 769, 155-167.
[46] Gao, S., You, J., Zheng, X., Wang, Y., Ren, R., Zhang, R., Bai, Y. and Zhang, H. (2010). *Talanta*, 82, 1371-1377.

[47] Tsoukali, H., Theodoridis, G., Raikos, N. and Grigoratou, I. (2005). *Journal of Chromatography* B, 822, 194-200.
[48] Farajzadeh, M. A., Djozan, D., Mogaddam, M. R. A. and Bamorowat, M. (2011). *Journal of Separation Science*, 34, 1309-1316.
[49] Lopez, R., Goni, F., Etxandia, A. and Millan, E. (2007). *Journal of Chromatography B*, 846, 298-305.
[50] Gallardo, E., Barroso, M., Margalho, C., Cruz, A., Vieira, D. and López-Rivadulla, M. (2006). *Analytical and Bioanalytical Chemistry*, 386, 1717-1726.
[51] Bielawski, D., Ostrea, E., Posecion, N., Corrion, M. and Seagraves, J. (2005). *Chromatographia*, 62, 623-629.
[52] Lacassie, E., Dreyfuss, M. F., Gaulier, J. M., Marquet, P., Daguet, J. L. and Lachâtre, G. (2001). *Journal of Chromatography B: Biomedical Sciences and Applications*, 759, 109-116.
[53] Barr, D. B., Leng, G., Berger-Preiss, E., Hoppe, H. W., Weerasekera, G., Gries, W., Gerling, S., Perez, J., Smith, K., Needham, L. L. and Angerer, J. (2007). *Analytical and Bioanalytical Chemistry*, 389, 811-818.
[54] Abu-Qare, A. W. and Abou-Donia, M. B. (2001). *Journal of Pharmaceutical and Biomedical Analysis*, 26, 291-299.
[55] Olsson, A., Nguyen, J., Sadowski, M. and Barr, D. (2003). *Analytical and Bioanalytical Chemistry*, 376, 808-815.
[56] Olsson, A. O., Baker, S. E., Nguyen, J. V., Romanoff, L. C., Udunka, S. O., Walker, R. D., Flemmen, K. L. and Barr, D. B. (2004). *Analytical Chemistry*, 76, 2453-2461.
[57] Hernández, F., Sancho, J. V. and Pozo, O. J. (2004). *Journal of Chromatography B*, 808, 229-239.
[58] Asensio-Ramos, M., Ravelo-Pérez, L. M., Gonzalez-Curbelo, M. A. and Hernandez-Borges, J. (2011). *Journal of Chromatography A*, 1218, 7415-7437.
[59] Jain, A. and Verma, K. K. (2011). *Analytica Chimica Acta*, 706, 37-65.
[60] Lee, J., Lee, H. K., Rasmussen, K. E. and Pedersen-Bjergaard, S. (2008). *Analytica Chimica Acta*, 624, 253-268.
[61] Zgola-Grzeskowiak, A. and Grzeskowiak, T. (2011). *TrAC Trends in Analytical Chemistry*, 30, 1382-1399.
[62] Han, D. and Row, K. (2012). *Microchimica Acta*, 176, 1-22.
[63] Tournier, M., Pouech, C., Quignot, N., Flament-Waton, M.-M., Lafay, F., Wiest, L., Lemazurier, E. and Cren-Olivé, C. (2012). *Analytical and Bioanalytical Chemistry*, submitted.
[64] Dulaurent, S., Moesch, C., Marquet, P., Gaulier, J.-M. and Lachâtre, G. (2010). *Analytical and Bioanalytical Chemistry*, 396, 2235-2249.
[65] Panuwet, P., Nguyen, J., Kuklenyik, P., Udunka, S., Needham, L. and Barr, D. (2008). *Analytical and Bioanalytical Chemistry*, 391, 1931-1939.
[66] Abdel-Rehim, M. (2010). *Journal of Chromatography A*, 1217, 2569-2580.
[67] Kumar, A., Gaurav, G., Malik, A. and Matysik, F.-M. (2009). *Bioanalytical Reviews,* 1, 35-55.
[68] Pragst, F. (2007). *Analytical and Bioanalytical Chemistry*, 388, 1393-1414.
[69] Kataoka, H. and Saito, K. (2011). *Journal of Pharmaceutical and Biomedical Analysis*, 54, 926-950.

[70] Lazartigues, A., Wiest, L., Baudot, R., Thomas, M., Feidt, C. and Cren-Olivé, C. (2011). *Analytical and Bioanalytical Chemistry*, 400, 2185-2193.
[71] Pouech, C., Tournier, M., Quignot, N., Kiss, A., Wiest, L., Lafay, F., Flament-Waton, M.-M., Lemazurier, E. and Cren-Olivé, C. (2012). *Analytical and Bioanalytical Chemistry*, in press.
[72] McClory, J., Henze, R., Tucker, K. and Thomas, R. (2009). *Lc Gc North America*, 27, 564-570.
[73] Winnik, B., Barr, D., Thiruchelvam, M., Montesano, M., Richfield, E. and Buckley, B. (2009). *Analytical and Bioanalytical Chemistry*, 395, 195-201.
[74] Chung, S. W. C. and Chen, B. L. S. (2011). *Journal of Chromatography A*, 1218, 5555-5567.
[75] Saito, K., Sjodin, A., Sandau, C. D., Davis, M. D., Nakazawa, H., Matsuki, Y. and Patterson Jr, D. G. (2004). *Chemosphere*, 57, 373-381.
[76] Anastassiades, M., Lehotay, S. J., Stajnbaher, D. and Schenck, F. J. (2003). *Journal of AOAC International*, 86, 412-431.
[77] Wilkowska, A. and Biziuk, M. (2011). Food Chemistry, 125, 803-812.
[78] Plossl, F., Giera, M. and Bracher, F. (2006). *Journal of Chromatography A*, 1135, 19-26.
[79] Hernández, F., Portoles, T., Pitarch, E. and Lopez, F. J. (2009). *Journal of Mass Spectrometry*, 44, 1-11.
[80] Tsai, W. H., Chuang, H. Y., Chen, H. H., Huang, J. J., Chen, H. C., Cheng, S. H. and Huang, T. C. (2009). *Analytica Chimica Acta*, 656, 56-62.
[81] Hajslová, J. and Zrostlikova, J. (2003). *Journal of Chromatography A*, 1000, 181-197.
[82] Anastassiades, M., Mastovská, K. and Lehotay, S. J. (2003). *J. Chromatogr. A*, 1015, 163-184.
[83] Gros, M., Petrovic, M. and Barceló, D. (2006). *Talanta*, 70, 678-690.
[84] Hewavitharana, A. K. (2011). *J. Chromatogr. A*, 1218, 359-361.
[85] Shabir, G. A. (2003). *J. Chromatogr. A*, 987, 57-66.
[86] Rambla-Alegre, M., Esteve-Romero, J. and Carda-Broch, S. (2012). *Journal of Chromatography A*, in press.
[87] ICH. (2005). Harmonised tripartite guideline Validation of analytical procedures: *text and methodology* Q2(R1).
[88] Max, F. (2007). *Journal of Chromatography A*, 1158, 174-183.
[89] AOAC. (1998). AOAC Peer-Verified Methods Program, *Manual on Policies and Procedures*, Arlington, VA, USA.
[90] Ebrahimi, M., Es'haghi, Z., Samadi, F. and Hosseini, M.-S. (2011). *Journal of Chromatography A*, 1218, 8313-8321.

In: Pesticides: Characteristics, Uses and Health Implications ISBN: 978-1-62081-577-9
Editors: A. Abrego Gomez and E. Lugo de Ortega

Chapter 3

IMPLICATIONS OF PESTICIDE USE IN OLIVE SECTOR: RESIDUE ANALYSIS

Antonio Ruiz-Medina * ***and Eulogio J. Llorent-Martínez***
Department of Physical and Analytical Chemistry,
Faculty of Experimental Sciences, University of Jaén,
Campus Las Lagunillas, Jaén, Spain

ABSTRACT

The application of phytosanitary products is one of the main employed procedures to solve the problems of plagues and diseases in olive groves. As a result, there is an increasing unease related to the possible presence of residues from these products in olive-derived foods. Recent studies have shown that these residues can be found in olives and olive oil (mainly insecticides, fungicides and herbicides) due to the lipophillic nature of the used plaguicides, usually at levels below the maximum residue limits (MRLs) established. Due to the fact that virgin olive oil production has increased in recent years and that it is being exported to different countries, exhaustive quality controls are required. In this sense, the quantification of pesticides residues is one of the most demanded analyses nowadays.

Olive farmers employ a wide range of pesticides in the plantations, sometimes without full understanding of their impact on human health and the environment. This constitutes a risk of these pesticides being incorporated in both: (1) rivers and reservoirs due to the soil that is eroded in heavy rains, polluting surface waters, etc.; and (2) olives and, consequently, olive oil and table olives. Different regulations regarding MRLs in table olives and olive oil have been established by the European Union and the Codex Alimentarius of the Food and Agriculture Organization of the United Nations. There are alternative methods available for the analysis of pesticides in different kinds of samples. However, the most common methods of analysis are Gas Chromatography (GC) and Liquid Chromatography (LC).

Our research group has performed thorough investigation over the analysis of pesticides in soils, olive fruits and virgin olive oils. In particular, we have developed different GC analytical methods with mass spectrometric detection for the analysis of all

* Tel: (+34)953-212759; Fax: (+34)953-212940; E-mail: anruiz@ujaen.es.

the main pesticides used in olive groves. In this way, sensitive and reliable procedures can be implemented for the quality control of table olives and olive oils in the industry. After describing the process for olive oil production, together with the common plagues and pesticides used in the olive groves, the methodologies developed by our research group will be described. Finally, the results obtained will be clearly explained.

1. Introduction

Olive oil is the main ingredient of the Mediterranean diet, due to its beneficial effects on human health. As a result, olive groves represent one of the main plantations in the countries of the Mediterranean Basin, especially in the South and Center of Spain and Italy, in Greece, Turkey, Tunisia and Morocco. The present situation of olive groves is the result of a slow and continuous development of the different civilizations that inhabited Mediterranean coastal areas and the inner lands of Middle East. The economic significance of olive oils varies among the producer countries and can be crucial in some of the mentioned countries, where the olive oil sector represents a high percentage of the economic activity. For instance, in Spain, more than half a million farmers work in olive groves. In other countries, such as Tunisia, olive oil is more than just goods; it is a strategic product for exportation activities, as the country has obtained valuable subsidizations from the European Union.

At first sight, olive oil does not seem to be so important when global statistical data in terms of production and consumption are observed. However, olive oil really represents a remarkable percentage of worldwide trading because its value per unit of product is significantly higher than that one of other edible oils (soybean, corn, sunflower, etc). The beneficial effects of consuming olive oil are backed by lengthy, painstaking scientific research. Olive oil contains a series of compounds that are very beneficial to most functions of the human body and its biological and therapeutic value is related in many aspects to its chemical structure, specially virgin olive oil, which is a natural product obtained from the olive fruit (*Olea Europaea*) using solely mechanical or physical means. The fact that oil extraction is solvent-free and natural antioxidants are maintained in the oil is reflected in the high nutritional and economic value of this product. Therefore, consumer's demand of this product has increased worldwide, not only in Mediterranean countries.

The most important plagues and diseases of the olive tree are the olive fruit fly, the olive moth and black scale. The application of phytosanitary products is one of the main employed procedures to solve these problems. It is important to understand the routes of elimination of the pesticides and monitoring their concentrations has become an important aspect of safeguarding the environment and public health. For these reasons, there is a need to develop sensitive and selective methods of analysis for the determination of these pesticides in agricultural plantations, such as olive groves. In addition, exhaustive quality control is required over the final products (table olives and olive oil) in order to ensure the absence of pesticides residues. The most important methods of analysis usually involve GC with mass spectrometry detection, as will be shown in this chapter.

2. Olive Oil Production

The technology for olive oil extraction, called elaiotechnic, is the fundamental basis for obtaining high-quality virgin olive oil (extra virgin). The main steps required for the elaboration of virgin olive oil follow.

Harvest of Olives

This step could be understood as an independent activity from the elaboration of the oil. However, the characteristics of the oil are highly influenced by the harvest time and method employed.

Olives must be picked at the moment of optimum ripeness, where the fruit presents the maximum content of oil and the best characteristics. Olives reach their ripeness in autumn and picking starts at the end of November, lasting up to February or March.

The methods to harvest the olives have not changed much from the ancient times; they should not damage the fruit and should avoid breaking of boughs or shoots. The high-quality olive oil is obtained by "milking" the olives into a sack tied around the harvester's waist (using ladders for the highest boughs) and extending canvases at the foot of the trees, where the olives will fall when the tree is beaten (using flexible poles). Recent methods make use of harvester machines (shakers) that generate the vibration on the tree for the falling of the fruit. This vibration machines can be small vibrators operated by the farmer in the required boughs or fully-equipped tractors, with vibration units for the whole trunk of the tree and an umbrella to pick the olives.

Harvest by hand is impossible in olive trees of 4-5 meters high, even with the use of ladders. In general, the olives are picked directly from the ground after they fall down in a natural way when they reach ripeness. When this happens, weeds should be completely eliminated from the zone in order to make it easier the picking. Anyway, the quality of the olive oil obtained from these olives is poor, because organoleptics flaws, such as soil flavor, are very evident.

Washing

Traditionally, agricultural workers have cleaned the fruit in the fields, by means of sieves. But this cleaning is not complete and the fruit goes into the olive-oil mill with a great quantity of impurities, leaves, branches, mud, etc., which is necessary to eliminate. Therefore, for this purpose, cleaners that use air current or shaking sieves are used to eliminate leaves, branches and other impurities lighter than the fruits. In addition, washing devices are employed to eliminate heavier impurities, such as stones or dust. In the next figure, both cleaning steps, the one with the air for light impurities and the other one by washing for heavy impurities, are shown.

Once the fruit has been cleaned and weighed, it is stored in hoppers until it can be crushed. The storage period must be as short as possible in order to avoid alterations, which would produce oils with a higher degree of acidity, lower stability and worse flavor.

Figure 1. Cleaning steps for olive fruits.

Preparation

The process of releasing the oil from the plant tissue begins by milling the olives to tear the flesh cells in order to let the oil run out of the vacuoles. This is followed by stirring the olive paste to permit the formation of large drops of oil and to break up the oil–water emulsion. In the so-called "dual-phase decanters" the oil is then separated by direct continuous centrifugation from the pomace, which consists of vegetable matter and water. The yield of oil varies from 80 to 90% of the total oil content of the olives, because the oil in the olive paste is only partially free to escape and part of it remains in the unbroken cells or is trapped in the tissues of the cytoplasm, or is emulsified in the aqueous phase.

After the extraction of virgin olive oil from the olives, the remaining paste is called pomace and still contains a small quantity (2-6%) of oil that can only be extracted with chemical solvents. This is done in specialized chemical plants, not in the olive oil mills and the obtained oil is called pomace oil.

The limit of FFA in olive oil for direct consumption is 2%. Virgin olive oils with higher content are called "lampantes" and must be refined prior to consumption. The components to be removed are all those ones that are detrimental to the flavor, color and stability of the oil, mainly FFA, phosphoacylglycerols, pigments, volatiles and contaminants. The standard processes used are chemical and physical refining. The main difference between both processes is that chemical refining procedure includes caustic soda treatment to neutralize the oil while, following physical refining, FFA are eliminated by distillation during deodorization. Physical refining reduces the loss of neutral oil, minimizes pollution and enables recovery of high quality FFA. Nevertheless, not all oils can be physically refined.

Taking into account all the procedure in order to obtain the oil, edible olive oil is marketed in accordance with the following designations and definitions [1]:

(a) *Virgin olive oil*: it is the oil obtained from the fruit of the olive tree solely by mechanical or other physical means under conditions, particularly thermal conditions, that do not lead to alterations in the oil, and which have not undergone any treatment other than washing, decantation, centrifugation and filtration. The virgin olive oils that are fit for consumption are extra virgin olive oil and virgin olive oil, presenting a free acidity, expressed as oleic acid, of not more than 0.8 and 2 grams per 100 grams, respectively. They also have to fulfil different quality criteria that, according to the

IOC, include good organoleptic characteristics (taste and aroma) and low peroxide value.

(b)*Olive oil*: it is the oil consisting of a blend (approximately 80:20 v:v) of refined olive oil and virgin olive oil fit for consumption as they are. It has a free acidity, expressed as oleic acid, of not more than 1 gram per 100 grams and its other characteristics correspond to those fixed for this category in this standard.

(c)*Olive pomace oil*: it is the oil obtained by treating olive pomace with solvents or other physical treatments, to the exclusion of oils obtained by re-esterification processes and of any mixture with oils of other kinds. The marketed olive pomace oil is the oil comprising the blend (approximately 80:20 v:v) of refined olive-pomace oil and virgin olive oil fit for consumption as they are. It has a free acidity (expressed as oleic acid) of not more than 1 gram per 100 grams.

Chemical processing may improve high acidity olive oil and make it edible; however, it takes away some extremely valuable ingredients such as vitamins and phenols. As a result, processed olive oil (refined) lacks the desirable properties and characteristics that can be found in abundance in virgin olive oil.

3. Plagues in Olive Groves

The attacks of pests and diseases and the presence of weeds make it necessary to apply pesticides to olive trees to ensure crop protection. The denomination pesticides (or plaguicides) include a wide variety of products that are very different in their chemical composition and characteristics. In general, plaguicides can be described as substances that control vegetable production and protect it from harmful organisms.

Taking into account the specific action of the plaguicides, different classifications can be made, being the decimal classification one of the most frequently used:

- Insecticides are used against insects and they include ovicides and larvicides (against eggs and larvae of insects, respectively). Nearly all insecticides have the potential to significantly alter ecosystems, being many of them toxic to humans.
- Acaricides kill members of the Acari group, which includes ticks and mites.
- Fungicides are chemical compounds or biological organisms used to kill or inhibit fungi or fungal spores.
- Nematocides, disinfectants and fumigants in general, used to kill parasitic nematodes.
- Herbicides are used to kill unwanted plants while leaving the desired crop relatively unharmed.
- Phytoregulators and similar products can be used to improve the potential of the trees.
- Molusquicides and rodenticides are used to control molluscs (slugs and snails) and rodents pests.
- Post-harvest pesticides and seeds.
- Protectors of woods, fibers and derivatives.

- Other specific plaguicides.

Laboratory studies show that pesticides can cause health problems, such as birth defects, nerve damage, cancer, and other effects that might occur over a long period of time. However, these effects depend on how toxic the pesticide is and how much of it is consumed. Some pesticides also pose unique health risks to children. For these reasons, the governments carefully regulate pesticides to ensure that their use does not pose unreasonable risks to human health or the environment.

The use of pesticides is the main way to fight against pests and plagues in olive groves. The plagues and diseases of the olive tree have changed little since ancient times. The most important ones are the olive fruit fly and the olive moth. The third one in importance is black scale, which appeared around the sixties. These plagues will be briefly described.

Olive Fruit Fly

The olive fruit fly *(Bactrocera oleae)* is a species of fruit fly which belongs to the Dacinae subfamily. It is a kind of phytophagous species, whose larvae feed on the fruit of olive trees, hence the common name. It is considered the most serious pest in the cultivation of olives. It is found throughout the Mediterranean Basin and in South Africa. Since the late 1990s it has also been present in California and has possibly spread throughout the area of olive cultivation in the Nearctic region.

The damage caused by the olive fruit fly can be quantitative and qualitative. In most cases, the greatest damage occurs as the fruit begins to soften and turn color, from September to November. The adult female can lay 50 to 400 eggs, generally one in each fruit. Olive fly eggs are small and difficult to see, embedded under the fruit surface. The tiny larvae hatching from the eggs are also quite difficult to see until they feed for a while and get larger. The first damage sign is an oviposition "sting" on the fruit surface. This looks like a small indentation or lump on the surface. Sometimes there is discoloration. While feeding, the larvae tunnel through the fruit, destroying the pulp and allowing entry of secondary infestation of bacteria and fungi that rot the fruit and greatly lower the quality of oil. The oil's FFA level ("acidity") increases considerably. If the damage is extensive enough, it may cause premature fruit drop. In areas of the world where the olive fruit fly is established and not controlled, its damage has been responsible for losses of up to 80% of oil value because of lower quantity and quality, and in some varieties of table olives, this pest is capable of destroying 100% of the crop.

The chemical battle against the olive fruit fly can be implemented against larvae by using preventive treatments against adults. The treatments are carried out by spraying the olive trees with insecticides, such as dimethoate, which is the most common due to its effectiveness and relatively low cost.

Olive Moth

Olive moth or olive kernel borer (*Prays oleae Bern) is* an important insect in the Mediterranean Basin and the United States and the second plague in importance. Three yearly generations are typical of this species. Each one is associated during the larval stage to different parts of the host. First generation larvae attack the flower buds, the second generation larvae grow inside the olive stone kernel and the third generation larvae are leaf mine. The damage (leaf drop) done by the leaf or third generation is seldom serious. On the contrary, damage caused by flower generation can be important. The larvae directly destroy the flowers or cause the abortion of the flower bunches covered by silken threads spun by the larvae when passing from one bud to the next. The fruit generation larvae cause the premature drop of the fruits when they bore the kernel of the olive fruit or later when they try to vacate the fruit to pupate. If the fall of the fruit takes place in June, the tree compensates this effect by increasing the size of the remaining fruits. However, a great reduction in the yielding and quality of the oil would be caused if the falling occurs in September.

There are two moments of action to fight this plague, at the beginning of the flowering (the period when the larvae are outside the fruit) or when the larvae are introducing themselves in the fruit. The plaguicides usually employed for the control of the olive moth are dimethoate and formothion, although others such as carbaryl can be used too.

Black Scale

Black scale (*Saissetia oleae*) is very common in all Mediterranean countries where olive tree is cultivated. This plague is caused by small black insects that cover the plants' leaves and stems. They sap the nutrients from the tree and excrete sticky, shiny honeydew on leaves of infested trees. At first, affected trees and leaves glisten and then become sooty and black in appearance as sooty mold fungus grows on the honeydew. Infestations reduce vigor and productivity of the tree, because the plant is denied the necessary sunlight. Continuous feeding reduces bloom the following year.

For controlling small amounts of black scale, it is possible to simply brush it off the stem or leaf with a toothbrush or knife. However, for larger infestations, chemical treatment could be required, although spraying oil on the plant can suffocate the pest sometimes. Some of the pesticides that can be used against black scale are dimethoate, diazinon and carbaryl.

4. Pesticides Employed in Olive Groves

As a result of the cited plagues, it is necessary to introduce the use of insecticides in the olive groves not only for the treatment of established plagues, but also for prevention. In addition, herbicides are required to control the growth of weeds. The absence of weeds in the soil around the olive tree presents two major benefits: the weeds do not waste water resources for the olive trees and the harvest of olives from the soil is easier when they naturally fall down from the tree.

In Figures 2-5, the pesticides that are usually employed in the Spanish olive groves are shown, classified by chemical families: organochlorines, orgahophosphorous, triazines and other families (phtalimides, carbamates and phenylureas).

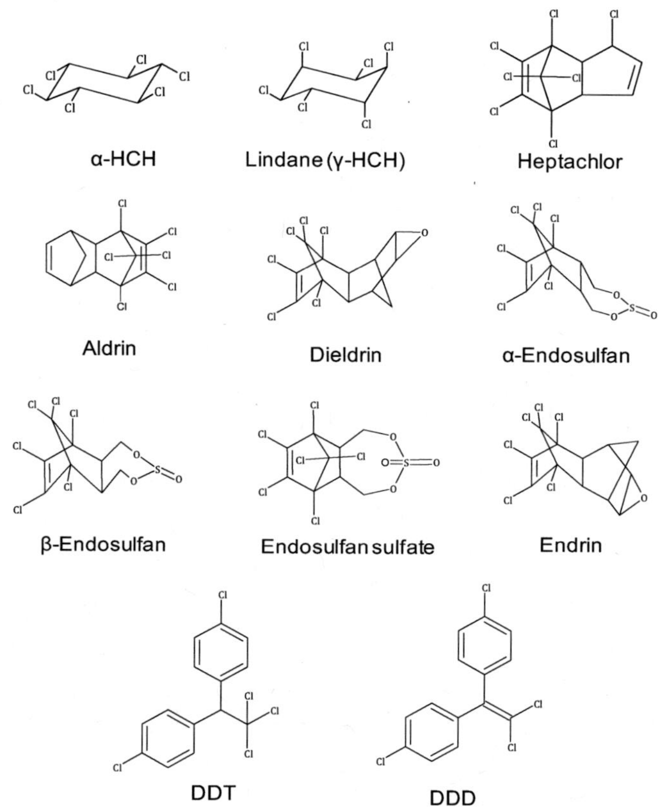

Figure 2. Organochlorines.

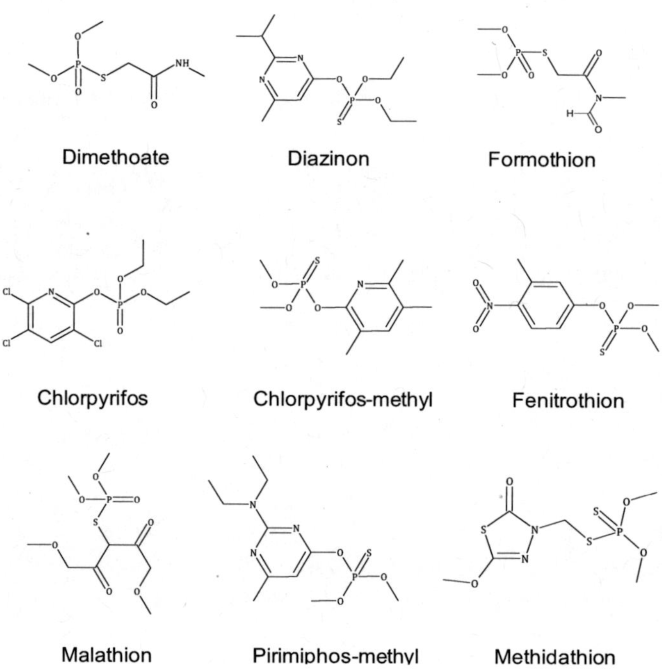

Figure 3. Orgahophosphorus.

In these Figures, not only plaguicides used in olive trees are shown, but also others that are employed in plantations close to olive groves and can be found in the olive oil. In addition, the use of some pesticides such as DDT or simazine is forbidden, but they are very persistent and they have to be analyzed in order to ensure the absence of their residues in the food.

The mechanisms of action of the plaguicides over the organism are very different depending on their chemical composition. These mechanisms are well-known for some of them, even at the molecular level. However, they are completely unknown for some others. Even among pesticides from the same family, some of them can be classified as scarcely toxic while others are very toxic. As a result, it is difficult to establish general rules when dealing with the toxicity of plaguicides. Organochlorine pesticides are insecticides with a highly stable, low volatile, lipophilic nature, consequently exhibiting considerable environmental persistence with a tendency to bioaccumulate, especially in foodstuffs with a high fat content. Many organochlorine compounds affect the central nervous system, the liver and the kidneys. They accumulate in adipose tissues and are slowly eliminated. With the introduction of new regulations against their use, other insecticides replaced them, such as the carbamate carbaryl and the organophosphorus, which are less stable in the environment. However, some of them affect the central nervous system, inhibiting the acetylcholinesterase enzyme.

Triazines, phenylurea diuron and phenylether oxyfluorphen are herbicides too, whereas phtalimides (captan, folpet) act as fungicides.

Monitoring pesticide residues in olive oil and olives is therefore of great interest to ensure food safety related to their use. The development of multi-residue methods, which allow proper control of a large number of pesticides in the same analysis, is required. These methods need to present a high sensitivity in order to be able to analyze the samples at the legislated MRLs.

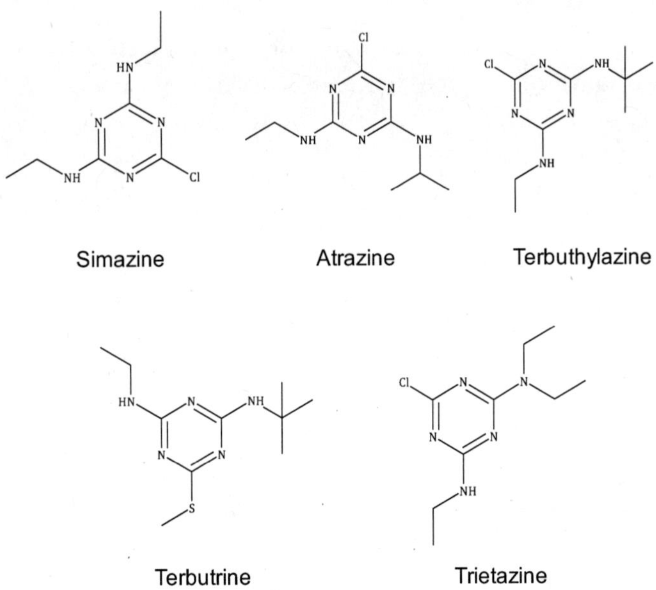

Figure 4. Triazines.

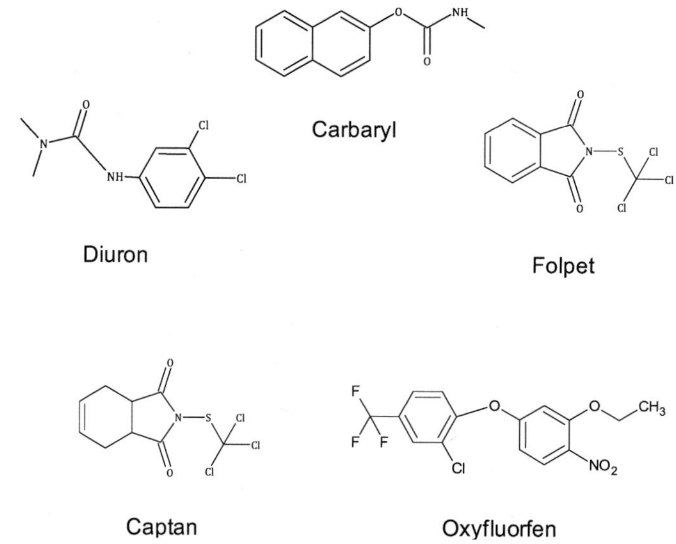

Figure 5. Other families.

The analysis of pesticide residues in olive oil is very challenging, because of the inherent complexity of the matrix. As a result, it is necessary to extract the pesticide fraction from the whole matrix to isolate the compounds that will be analyzed. Taking into account that some pesticides are fat-soluble, it is difficult to completely separate them from the matrix. Hence, a clean-up step is required after the extraction procedure.

5. Gas Chromatography as Detection Technique

Before pesticides can be determined in any complex sample, such as olives or olive oils, they have to be separated from the matrix and other interfering compounds, in order to isolate them from the rest of the sample. This separation includes extraction, preconcentration and clean-up steps [2,3].

This part of the analysis is critical, because the final results will be completely related to the success of the separation process. The sample has to come into contact with an extractant (solid, liquid or supercritical fluid) in optimized conditions in order to weaken the analyte-matrix interactions and increase the analyte-extractant interactions.

There is a continuous development of sample-treatment procedures for the isolation of pesticides in samples with relatively high fat content, such as olive oil and olives. The preparation of oil samples for their determination by chromatographic techniques requires the removal of the fatty components from the sample. The main problem associated when working with this kind of samples is that dirty extracts may harm the instruments employed. For this reason, proper extraction and clean-up steps are required. In addition, the different nature and physicochemical properties of the classes of pesticides to be analyzed makes it more complex to select the appropriate extraction methodology.

A data that is usually employed for each pesticide is the $K_{O/W}$ value, which is the partition coefficient between octanol and water, and its logarithm is an indication of fat solubility of the plaguicide. A high value of the coefficient $K_{O/W}$ for a particular plaguicide indicates that its solubility in water is low; hence, this plaguicide would be fat-soluble and there would be risks of bio-accumulation in fatty tissues. It has been demonstrated that fat-soluble pesticides tend to concentrate in olive oil during its production and extraction from the olives. For this reason, higher concentration levels of fat-soluble plaguicides are expected in the oil than in the olives from which the oil was produced. On the other hand, polar pesticides do not tend to preconcentrate in olive oil and their concentration is lower.

There are numerous extraction and clean-up procedures. However, the most common ones are liquid-liquid or solid-liquid extraction, solid-phase extraction (SPE) and gel permeation chromatography (GPC) [3-5].

Liquid-Liquid Extraction

Usually, it has been the chosen method for the extraction, preconcentration and clean-up of liquid samples. It is based on the relative solubility of the analyte in two different immiscible liquids; therefore, it is an extraction of a substance from one liquid phase into another liquid phase. The liquid sample comes to contact with an appropriate solvent (immiscible with the sample) and, after a mixing time, both phases (sample and extractant) are separated. During the mixing period, the analyte is distributed between both phases until it reaches the equilibrium.

The $K_{O/W}$ value is critical to decide the appropriate extractant. If this value is high, it indicates that the analyte tends to dissolve in the organic phase instead of in water. As a general rule, the organic solvents selected are those one that present high affinity for the analytes, are immiscible with the sample and volatile (to evaporate them easily and preconcentrate the analytes).

Solid-Liquid Extraction

In this case, the sample comes to contact with an appropriate solvent. After that, different procedures can be applied to homogenize both phases: ultrasonic agitation and microwave extraction.

Ultrasonic agitation: the solvent interacts with the sample simply by a shaking process. Ultrasonic radiation causes the vibration of the molecules, increasing the collision between them. Hence, the contact between sample and solvent is enhanced. After the analyte is dissolved in the extractant, a filtration step and the evaporation of the solvent are required.

Microwave extraction: samples are enclosed in high quality Teflon vessels together with the solvent and heated to a controlled temperature with microwave power, being the extract filtered when the process finishes. Electromagnetic irradiation is used to heat the solvent that acts as extractant, which is in most cases a mixture between hexane and acetone.

Solid-Phase Extraction

It is a separation process by which compounds that are dissolved or suspended in a liquid mixture are separated from other compounds in the mixture according to their physical and chemical properties. SPE is used to concentrate and purify samples for analysis and to isolate the analytes of interest from a wide variety of matrices.

SPE uses the affinity of the analytes dissolved or suspended in a liquid (known as the mobile phase) for a solid through which the sample is passed (known as the stationary phase) to separate a mixture into desired and undesired components. The result is that either the desired analytes of interest or undesired impurities in the sample are retained on the stationary phase. The portion that passes through the stationary phase is collected or discarded, depending on whether it contains the desired analytes or undesired impurities. If the portion retained on the stationary phase includes the desired analytes, they can then be removed from the stationary phase for collection in an additional step, in which the stationary phase is rinsed with an appropriate eluent.

The stationary phase comes in the form of a packed syringe-shaped cartridge, each of which can be mounted on its specific type of extraction manifold. The manifold allows multiple samples to be processed by holding several SPE media in place and allowing for an equal number of samples to pass through them simultaneously. A typical cartridge SPE manifold can accommodate up to 24 cartridges and is equipped with a vacuum port. Application of vacuum speeds up the extraction process by pulling the liquid sample through the stationary phase. The analytes are collected in sample tubes inside or below the manifold after they pass through the stationary phase.

Solid phase extraction cartridges are available with a variety of stationary phases, each of which can separate analytes according to different chemical properties. Most stationary phases are based on silica that has been bonded to a specific functional group.

Gel-Permeation Chromatography

It is probably the most extensively used technique for the analysis of pesticide residues in olive oil, usually after a liquid-liquid extraction. In this technique, an aliquot of an olive oil extract (obtained from the extraction step) is injected into the GPC system. The selected fraction is collected and, after a solvent-exchange step, the sample is analyzed by gas chromatography. A GPC system is composed of a chromatographic pump, a fraction collector and a detector. The columns are made from polymeric porous microspheres that enable the separation of compounds according to their molecular weights (which are related to the size of the compound). Using this principle, the pesticide fraction is separated from the triglyceride fraction, which presents higher molecular weight. In GPC, the compounds are eluted from higher to lower molecular weight.

Gas Chromatography

Gas chromatography (GC) is an analytical technique widely used for separating and analysing compounds that can be vaporized without decomposition [6,7]. Typical uses of GC include testing the purity of a particular substance, or separating the different components of a mixture.

In GC, the mobile phase is a carrier gas, usually an inert gas such as helium or an unreactive gas such as nitrogen. The stationary phase is a microscopic layer of liquid or polymer on an inert solid support, inside a piece of glass or metal tubing called column. The instrument used to perform gas chromatography is called a gas chromatograph.

A gas chromatograph uses a flow-through narrow tube known as the column, through which different chemical constituents of a sample pass in a gas stream (carrier gas, mobile phase) at different rates depending on their various chemical and physical properties and their interaction with a specific column filling, called the stationary phase. As the chemicals exit at the end of the column, they are detected and identified electronically. The function of the stationary phase in the column is to separate different components, causing each one to exit the column at a different time (retention time). Other parameters that can be used to alter the order or time of retention are the carrier gas flow rate, column length and the temperature.

In a GC analysis, a known volume of gaseous or liquid sample is injected into the head of the column using a microsyringe. As the carrier gas sweeps the analyte molecules through the column, this motion is inhibited by the adsorption of the analyte molecules either onto the column walls or onto packing materials in the column. The rate at which the molecules progress along the column depends on the strength of adsorption, which in turn depends on the type of molecule and on the stationary phase materials. Since each type of molecule has a different rate of progression, the various components of the mixture of analytes are separated as they progress along the column and reach the end of the column at different times. A detector is used to monitor the outlet stream from the column; thus, the time at which each component reaches the outlet and the amount of that component can be determined.

The column(s) in a GC are contained in an oven, the temperature of which is precisely controlled electronically. The rate at which a sample passes through the column is directly proportional to the temperature of the column. The higher the column temperature, the faster the sample moves through the column. However, the faster a sample moves through the column, the less it interacts with the stationary phase, and the less the analytes are separated. Two types of columns are used in GC:

- Packed columns are 1.5–10 m in length and have an internal diameter of 2–4 mm. The tubing is usually made of stainless steel or glass and contains a packing of finely divided, inert, solid support material (e.g. diatomaceous earth) that is coated with a liquid or solid stationary phase. The nature of the coating material determines what type of materials will be most strongly adsorbed.
- Capillary columns have a very small internal diameter, in the order of a few tenths of millimeters, and lengths between 25–60 meters are common. Most capillary columns are made of fused-silica with a polyimide outer coating. These columns are flexible, so a very long column can be wound into a small coil.

The most common detectors in GC until recent years were the flame ionization detector (FID), the thermal conductivity detector (TCD) and the electron capture detector (ECD). However, nowadays most of the developed analytical methods use a mass spectrometry (MS) detector.

MS is an analytical technique that measures the mass-to-charge ratio of charged particles. It is used for determining masses of particles, for determining the elemental composition of a sample or molecule, and for elucidating the chemical structures of molecules. The MS principle consists of ionizing chemical compounds to generate charged molecules or molecule fragments and measuring their mass-to-charge ratios.

In a typical MS procedure:

1. A sample is loaded onto the MS instrument, and undergoes vaporization.
2. The components of the sample are ionized by one of a variety of methods (e.g., by impacting them with an electron beam), which results in the formation of charged particles (ions).
3. Ions are separated according to their mass-to-charge ratio in an analyzer by electromagnetic fields.
4. Ions are detected, usually by a quantitative method.
5. Ion signal is processed into mass spectra.

MS instruments consist of three modules: a) an ion source, which produces ions from analytes present in the sample; b) a mass analyzer, which sorts the ions by their masses; c) a detector that provides data for calculating the abundances of each ion present.

The main ion sources are electron ionization (EI) and chemical ionization (CI). EI is an ionization method in which energetic electrons interact with gas phase atoms or molecules to produce ions. The electrons are produced through termoionic emission by heating a wire filament that has electric current through it. The electrons are accelerated and concentrated into a beam; once the sample is introduced into the ion source, the neutral molecules collide with the electron beam and the analyte molecules are ionized. In CI, the ions are produced through the collision of the analyte molecules with ions of a reagent gas, such as methane or ammonia. This gas is present in large excess compared to the analyte.

Although there are different mass analyzers, the most common one is the *quadrupole.* It consists of four parallel metal rods. Each opposing rod pair is connected together electrically, and a radio frequency (RF) voltage is applied between one pair of rods and the other. A direct current voltage is then superimposed on the RF voltage. Ions travel down the quadrupole between the rods. Only ions of an m/z ratio will reach the detector for a given ratio of voltages: other ions have unstable trajectories and will collide with the rods. This permits selection of an ion with a particular *m/z* or allows the operator to scan for a range of m/z-values by continuously varying the applied voltage. A linear series of three quadrupoles can be used; known as a *triple quadrupole* mass spectrometer. The first (Q_1) and third (Q_3) quadrupoles act as mass filters, and the middle (Q_2) quadrupole is employed as a collision cell. This collision cell is an RF-only quadrupole (non-mass filtering) using argon, hellium or nitrogen gas for collision induced dissociation of selected ions from Q_1. Subsequent fragments are passed through to Q_3 where they may be filtered or fully scanned.

The *quadrupole ion trap*, used in the methods described in this chapter, works on the same physical principles as the quadrupole mass analyzer, but the ions are trapped in the instrument for the required time and then ejected. Once the ions are produced in the source of the instrument, they enter through the inlet and are trapped through action of three hyperbolic electrodes: the ring electrode and the entrance and exit endcap electrodes. Various voltages are applied to these electrodes, resulting in the formation of a cavity in which ions are trapped in a stable oscillating trajectory. The exact motion of the ions is dependent on the voltages applied and their individual mass-to-charge (*m/z*) ratios. For detection of the ions, the potentials are altered to destabilize the ion motions resulting in ejection of the ions through the exit endcap. The very nature of trapping and ejection makes a quadrupolar ion trap especially suited to performing MS experiments in structural elucidation studies. It is possible to selectively isolate a particular *m/z* in the trap by ejecting all the other ions from the trap, and perform further fragmentation of this isolated precursor ion. The isolation and fragmentation steps can be repeated a number of times depending on the efficiency of the instrument.

6. QUANTITATION OF PESTICIDES

For the production of 1 L of olive oil, approximately 5 kg of olives are required. Taking into account that most pesticides are lipophilic, a concentration effect could occur when obtaining the olive oil. Therefore, MRL for pesticides have been set by the European Union in both olives and olive oil. For this reason, our research group has developed analytical methods that allow the analysis of the selected plaguicides in olive oil [8] and olives [9] by GC-MS.

Table 1. Parameters of the GC-MS system for the selected pesticides in olive oil

No.	Plaguicide	Linear Range (μg kg^{-1})	DL (μg kg^{-1})
1	Diuron	5-1000	0.5
2	α-HCH	1-1000	0.5
3	Dimethoate	20-400	10
4	Simazine	3-2000	1
5	Atrazine	5-1000	2
6	Trietazine	10-800	5
7	β-HCH	0.5-1000	0.2
8	Terbuthylazine	1-1000	0.5
9	δ-HCH	1-1000	0.5
10	Diazinon	5-1000	2
11	Lindane	0.5-1000	0.2
12	Formothion	20-1000	10
13	Chlorpyrifos-methyl	5-2000	1
14	Heptachlor	3-1000	1
15	Carbaryl	5-2000	1
16	Pirimifos-methyl	1-2000	0.5
17	Terbutrine	20-1000	5
18	Fenitrothion	3-2000	1
19	Chlorpyrifos	0.5-1500	0.2
20	Aldrin	10-1000	5
21	Heptachlor-epox-a	5-2000	1
22	Heptachlor-epox-b	5-2000	1
23	Folpet	10-2000	5
24	α-Endosulfan	5-2000	1
25	Dieldrin	5-2000	1
26	Endrin	10-2000	5
27	β-Endosulfan	5-2000	1
28	DDD	0.5-2000	0.2
29	Endosulfan sulfate	5-1000	2
30	DDT	5-2000	1

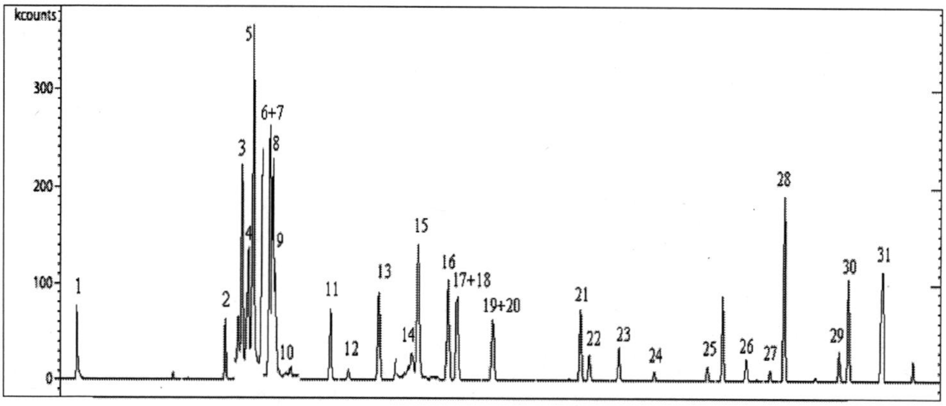

Figure 6. GC-MS chromatogram from a standard of pesticides in olive oil.

Table 2. Parameters of the GC-MS system for the selected pesticides in olives

No.	Plaguicide	Linear Range (μg kg^{-1})	DL (μg kg^{-1})
1	Diuron	1.25-250	0.12
2	α-HCH	0.25-250	0.12
3	Dimethoate	5-100	2.50
4	Simazine	0.75-125	0.25
5	Atrazine	1.25-250	0.5
6	Trietazine	2.50-200	1.25
7	β-HCH	0.12-250	0.05
8	Terbuthylazine	0.25-250	0.12
9	δ-HCH	0.25-250	0.12
10	Diazinon	1.25-250	0.5
11	Lindane	0.12-250	0.05
12	Formothion	5-250	2.5
13	Chlorpyrifos-methyl	1.25-500	0.25
14	Heptachlor	0.75-250	0.25
15	Carbaryl	1.25-500	0.25
16	Pirimifos-methyl	0.25-500	0.12
17	Terbutrine	5-250	1.25
18	Fenitrothion	0.75-500	0.25
19	Chlorpyrifos	0.12-375	0.05
20	Aldrin	2.5-250	1.25
21	Heptachlor-epox-a	1.25-500	0.25
22	Heptachlor-epox-b	1.25-500	0.25
23	Folpet	10-2000	5
24	α-Endosulfan	5-2000	1
25	Dieldrin	5-2000	1
26	Endrin	10-2000	5
27	β-Endosulfan	5-2000	1
28	DDD	0.5-2000	0.2
29	Endosulfan sulfate	5-1000	2
30	DDT	5-2000	1

The preparation procedure for olive oil samples included a liquid-liquid extraction followed by GPC clean-up step. The detection of the pesticides was performed making use of GC coupled to mass spectrometry. The following Table shows the analytical parameters obtained for each one of the 30 selected pesticides. DL stands for detection limit. The total separation of all the pesticides can be seen in the chromatogram shown in Figure 6. The numbers of the peaks correspond to those ones assigned to the pesticides in Table 1; peak number 31 corresponds to the internal standard triphenylphosphate.

For olive samples, approximately 130 g of olives (including the seeds) were first crushed by means of a hammer mill. Afterwards, a 100 g portion was weighed in a glass tube and 50 g of anhydrous sodium sulphate were added. The sample was then extracted with an adequate organic solvent and a GPC-based clean-up procedure was applied to the fatty residue obtained after the extraction of the pesticides from the olives. Finally, their detection was performed by means of GC-MS. In a similar way to olive oil, the analytical parameters of the system for olives are shown in Table 2.

The procedures previously detailed were applied to the determination of the cited pesticides in olives and olive oil samples. In Table 3, the fourteen detected pesticides in real samples are shown. In particular, the herbicides diuron, terbuthylazine and endosulfan sulfate were found in many samples. Although other pesticides were also detected, in most cases the levels were below the quantification limit of the system.

Influence of Harvest

Depending on the harvest method, the olive fruits can be grouped into three categories: a) fruits picked directly from the tree without any contact with the soil (flight olives); b) fruits picked from the ground (soil olives); and c) fruits that are not separated before the elaboration process (non-separated olives). The separation of flight and soil olive fruits before olive oil elaboration is critical in order to obtain appropriate results (high-quality virgin olive oil). If both fruits are not separated, the quality of the oil and the percentage of extra virgin olive oil obtained diminish. In addition, the harvesting method may be very important for the presence of pesticide residues in olives, especially herbicides that are applied to the soil.

Data from Table 3 show that herbicide diuron was quantified in a high number of samples in soil olives and olive oil. In four of them, the levels were higher than the MRL established by the European Union (0.2 and 0.8 mg kg^{-1}, respectively). Terbuthylazine and endosulfan sulfate were also quantified in many samples with values over the MRL in some of them, being the established MRLs 0.05 and 0.2 mg kg^{-1} for both pesticides in olives and olive oil, respectively.

These results suggest that pesticide residues are mainly caused by the contamination of the olives when they come to contact with the soil after falling down [8,10]. The levels found in soil olives were significantly higher than those ones found in flight olives, which had not been in contact with the soil. However, the levels of endosulfan sulfate were very similar in soil and flight olives. The reason is that this pesticide is applied by spray in the atmosphere, and not on the soil. Hence, it can be carried large distances through the air and the most plausible way of contamination is when the olives are still on the tree. This mechanism of contamination would explain some cases in which this pesticide was detected in olives even without being directly applied to the tree. In one olive mill in which the fruits were not separated before the olive oil elaboration, low levels of pesticides were observed, finding only one sample with terbuthylazine levels higher than its MRL.

Table 3. Number of samples with pesticides in olive and olive oil samples

Plaguicide	Olives			Olive oil		
	Soil	Flight	Non-separated	Soil	Flight	Non-separated
	n=50	n=20	n=24	n=15	n=10	n=8
Diuron	46	13	20	15	10	6
α-HCH	6	2	4	0	0	1
Simazine	2	1	1	5	2	2
Dimethoate	2	0	0	0	0	0
Atrazine	5	0	1	0	0	0
β-HCH	6	2	1	0	0	0
Terbuthylazine	42	13	24	15	10	7
Lindane	2	1	2	0	0	0
Carbaryl	2	1	3	0	0	0
Chlorpyrifos	2	0	0	5	2	2
α-Endosulfan	7	10	2	2	2	0
Dieldrin	2	0	0	0	0	0
β-Endosulfan	2	0	0	0	0	0
Endosulfan sulfate	26	9	16	9	9	5

n = total number of samples analyzed.

Influence of Washing in Olive Mill

Taking into account the few bibliographic data about the fate of pesticides after their application on the olive groves, it would be interesting to evaluate which fraction of the pesticides could be eliminated from the olives after the washing process in the olive mill. Therefore, a group of pesticides which frequently appeared in the analyzed samples were selected: diuron, terbuthylazine and endosulfan sulfate.

The fruit goes into the olive-oil mill with a great quantity of impurities, leaves, branches, mud, etc, which are necessary to eliminate. Therefore, for this purpose, cleaners and washing devices are employed to eliminate impurities, especially present in soil olives. Some olive mills separate flight and soil olives, whereas others treat all the fruits together, without any separation process. Our research group presented the first exhaustive study of the influence of the olives washing in the mills over the concentration of pesticides [11-13]. The obtained results in flight, soil and non-separated olives after the washing process are shown in Figure 7, where the total concentration of the three selected pesticides is represented for 47 samples (non-washed and washed olives). The most outstanding conclusion is the reduction in the levels of plaguicides in soil olives after the washing process when compared to the same olive samples before the washing step. Only two samples presented levels higher than the legislated MRL for the herbicide terbuthylazine.

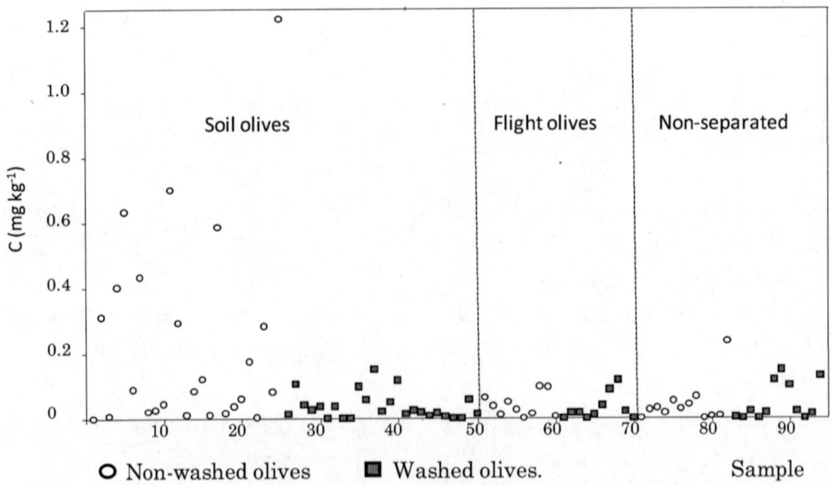

Figure 7. Concentration of pesticides in washed and non-washed olives.

It can be observed that the washing process significantly diminishes the levels of residues of plaguicides in soil olives, while the influence of the washing step is not clearly appreciated in flight olives. In the case of non-separated olives, it is interesting to remark that some of the washed olives present higher levels of pesticides than the non-washed ones. This can be due to the contamination of pesticides-free olives during the washing process. The washing machines for the olives are cleaned and filled with fresh water at the starting of the day. With this water, up to 160000 kg of olives can be washed before the water is changed next day. The water can be contaminated after the washing of soil olives, therefore contaminating following pesticides-free olives in the washing process. The washing waters (before and after the washing process) from the olive mill were analyzed and the obtained results confirmed this theory [13].

CONCLUSION

Virgin olive oil production and export has increased in recent years. As a result, it represents an important part of the economy in some Mediterranean countries. Although olive oil presents a lot of beneficial effects, it is necessary to carry out exhaustive quality controls in order to maintain its high standards. One of the most important ones is the analysis of residues of plaguicides, due to their toxicity, persistence, mobility and environmental accumulation. With prolonged use, there is a risk that these chemicals will gather in crops and soils and, later, leach into surface and ground waters. In this chapter, we have described the most common procedures developed for the analysis of pesticides in olives and olive oil. It has been shown that some of them are commonly found in these samples, such as diuron or terbuthylazine, although usually at levels below the established MRLs. In addition, the separation of flight and soil olives, together with the washing process in the mills, is a critical step in order to control the levels of herbicides. Hence, the exhaustive analyses that are routinely carried out for pesticides should be kept in order to continue ensuring the safety of the virgin olive oil.

REFERENCES

[1] International Olive Council (2010). *Trade Standard Applying to Olive Oils and Olive-Pomace Oils*. COI/T.15/NC,

[2] Fuentes, E., Báez, M.E., and Díaz, J. (2009). Microwave-assisted extraction at atmospheric pressure coupled to different clean-up methods for the determination of organophosphorus pesticides in olive and avocado oil. *Journal of Chromatography* A, 1216, 8859-8866.

[3] Gilbert-López, B., García-Reyes, J.F., and Molina-Díaz, A. (2009). Sample treatment and determination of pesticide residues in fatty vegetable matrices: A review. *Talanta,* 79, 109-128.

[4] Mukherjee, I. and Gopal, M. (1996). Chromatographic techniques in the analysis of organochlorine pesticide residues. *Journal of Chromatography* A, 754, 33-42.

[5] Walters, S.M. (1990). Clean-up techniques for pesticides in fatty foods. *Analytica Chimica Acta*, 236, 77-82.

[6] Harris, D.C. (2007): *Quantitative Chemical Analysis*. Freeman and Company, New York.

[7] Skoog, D.A., West, D.M., and Holler, F.J. (1996). *Fundamentals of Analytical Chemistry.* Saunders College, Philadelphia.

[8] Guardia-Rubio, M., Fernández-de Córdova, M.L., Ayora-Cañada, M.J., and Ruiz-Medina, A. (2006). Simplified pesticide multiresidue analysis in virgin olive oil by gas chromatography with thermoionic specific, electron-capture and mass spectrometric detection. *Journal of Chromatography* A, 1108, 231-239.

[9] Guardia-Rubio, M., Marchal-López, R.M., Ayora-Cañada, M.J., and Ruiz-Medina, A. (2007). Determination of pesticides in olives by gas chromatography using different detection systems. *Journal of Chromatography* A, 1145, 195-203.

[10] Guardia-Rubio, M., Ruiz-Medina, A., Molina-Díaz, A., and Ayora-Cañada, M.J. (2006). Influence of harvesting method and washing on the presence of pesticide residues in olives and olive oil. *Journal of Agricultural and Food Chemistry*, 54, 8538-8544.

[11] Guardia Rubio, M., Ruiz Medina, A., Pascual Reguera, M.I., and Fernández de Córdova, M.L. (2007). Multiresidue analysis of three groups of pesticides in washing waters from olive processing by solid-phase extraction-gas chromatography with electron capture and thermionic specific detection. *Microchemical Journal*, 85, 257-264.

[12] Guardia-Rubio, M., Ayora-Cañada, M.J., and Ruiz-Medina, A. (2007). Effect of washing on pesticide residues in olives. *Journal of Food Science*, 72, C139-C143.

[13] Guardia-Rubio, M., Ruiz-Medina, A., Molina-Díaz, A., and Ayora-Cañada, M.J. (2008). Pesticide residues in washing water of olive oil mills: Effect on olive washing efficiency and decontamination proposal. *Journal of the Science of Food and Agriculture*, 88, 2467-2473.

In: Pesticides: Characteristics, Uses and Health Implications ISBN: 978-1-62081-577-9
Editors: A. Abrego Gomez and E. Lugo de Ortega

Chapter 4

PESTICIDES AND ENDANGERED SEA TURTLES: A REVIEW

Jorge Orós[1], María Camacho[1], Luis D. Boada[2] and Octavio P. Luzardo[2]
[1]Veterinary Faculty, University of Las Palmas de Gran Canaria, Spain
[2]Toxicology Unit, Department of Clinical Sciences, University of Las Palmas de Gran Canaria, Spain

ABSTRACT

Sea turtles are considered of increasing interest as potential bioindicators for pollution in marine ecosystems. In addition, all species of sea turtles are included on the Red List of the World Conservation Union, and all the threats for their survival must be investigated, including the health implications of pesticides.

This chapter is a review of the international research made on pesticides and sea turtles. The majority of the studies reported baseline levels of organochlorine (OC) pesticides in several species of sea turtles from different geographical areas. Several sea turtle samples were used to determine the concentration of OC pesticides: eggs and chorioallantoic membranes, tissues, and blood. Contamination of sea turtle eggs by pesticides, effects on embryonic development, maternal transfer, and even implications for public health (consumption of sea turtle eggs) were investigated. Several specific studies investigated the effects of pesticides on sexual differentiation, sexual steroid-binding proteins, and immunity. Other relevant studies tried to establish an association between OC pesticides and clinical health parameters and/or diseases.

INTRODUCTION

Pesticides are well known as environmental contaminants because of their stability, bioaccumulative capacity and global occurrence. OC pesticides, such as DDT and its metabolites, have contaminated the global environment (Iwata *et al.* 1994). Once extensively used as domestic and agriculture pesticides (Ecobichon 1995), most of these compounds were banned from use in developed countries starting in the 1970s mainly because of their detrimental effects on wildlife and concerns regarding human health effects (Keller *et al.* 2004b). However, OC contaminants continue to be detected in environmental samples and animal tissues due to their long residual life and persistence, their ability to bioaccumulate through the food chain, and their continued use in some underdeveloped countries (Iwata *et al.* 1994, Bolt and Degen 2002, Keller *et al.* 2004b).

Two families and seven species of sea turtles are currently recognized (Pritchard 1997). The family Cheloniidae includes the green turtle (*Chelonia mydas*), loggerhead (*Caretta caretta*), hawksbill (*Eretmochelys imbricata*), Kemp's ridley (*Lepidochelys kempi*), olive ridley (*Lepidochelys olivacea*), and flatback turtle (*Natator depressus*). The family Dermochelyidae includes only the leatherback (*Dermochelys coriacea*).

All species of sea turtles are included on the Red List of the World Conservation Union (IUCN/SCC 2012). Pollution and pollution-related disease increasingly threaten turtle populations (Gardner *et al.* 2003). The IUCN emphasized as a priority issue in its document entitled *A Global Strategy for the Conservation of Marine Turtles* (1995) the research on the effects of persistent organic pollutants on sea turtles. Thus in recent years, increased efforts have been devoted to the conservation of sea turtles, including the health implications of pesticides.

Sea turtles are long-living vertebrates that may bioaccumulate these contaminants from food, sediment, and water. Thus sea turtles are considered of increasing interest as potential bioindicators for pollution in marine ecosystems (D'Ilio *et al.* 2011).In addition, the life history of sea turtles creates a challenge for toxicological research, because their long migrations facilitate their exposure to diverse contaminant classes arising from a variety of sources (Keller *et al.* 2004a).

This chapter is a review of the international research made on pesticides in sea turtles, including baseline levels of OC pesticides in different samples of several sea turtles species from different geographical areas, contamination of sea turtle eggs by pesticides, effects of pesticides on sexual differentiation, maternal transfer of pesticides, and effects of pesticides on immunity, clinical health parameters and/or diseases of sea turtles.

Baseline Levels of Organochlorine Pesticides in Sea Turtles

Several studies were carried out to determine the baseline levels of OC pesticides in sea turtles from different geographical areas (Pugh and Becker 2001). The majority of surveys describe values for DDT and its isomers in organs and tissues of sea turtles (Table 1). Other OC pesticides, such as chlordanes, mirex, endrin, dieldrin, lindane, endosulfan, hexachlorocyclohexane (HCHs), hexachlorobenzene (HCB), and toxaphene have also been reported in organs and tissues of sea turtles.

Table 1. DDT concentrations (average concentration or range, ng/g w. w.) in sea turtles

Species	Locality	Tissue (n)	∑DDT	*o,p'*-DDT	*o,p'*-DDD	*o,p'*-DDE	*p,p'*-DDT	*p,p'*-DDD	*p,p'*-DDE	Source
C. caretta	Florida	Muscle (9)							1-45	McKim and Johnson (1983)
		Liver (9)							2-100	
C. mydas		Muscle (4)							< 1	
		Liver (4)							<10	
C. mydas	Hawaii	Fat (10)		BDL	BDL	BDL	BDL	BDL	BDL	Aguirre *et al.* (1994)
L. kempi	Long Island	Liver (22)							29.8-547	Lake *et al.* (1994)
		Fat (13)							64.1-720	
C. caretta	Virginia and North Carolina	Liver (18)					BDL-3.44	BDL-4.08	BDL-458	Rybitski *et al.* (1995)
		Fat (20)					BDL-10.8	BDL-35.6	2.86-1210	
L. kempi		Liver (3)					BDL	4.94-11.4	54.2-56.8	
		Fat (3)					BDL-5.19	15.5-42.1	95.7-292	
D. coriacea	Britain	Fat (3)					<1	<1	10-68	Godley *et al.* (1998)
C. caretta	Med. Sea	Liver (4)	49-77				<0.5-<0.7	<0.2-<0.5	49-76	Mckenzie *et al.* (1999)
		Fat (3)	391-739				<2-8.0	<1.8-6.6	376-705	
C. mydas		Liver (9)	1-23				<0.4-<1.2	<0.4-<1.2	<1-23	
		Fat (3)	3.3-23				<1.9-<3.2	<1.9-<3.2	2.4-19	
C. caretta	Scotland	Liver (1)	152				-	-	149	
D. coriacea	Scotland	Liver (2)	9.1-14				<0.6-<0.7	<0.6-<0.7	1.7-6.5	
		Fat (2)	14-58				<3.2-<6.2	<6.2-<8.1	10-57	
C. caretta	Adriatic Sea	Liver (11)		BDL	BDL	-	BDL	BDL	1.42± 0.25[a] 0.41± 0.25[b]	Storelli and M. (2000)[*]
		Lung (11)		BDL	BDL	-	BDL	BDL	0.48± 0.24[a] 0.14± 0.08[b]	
		Kidney (11)		BDL	BDL	-	BDL	BDL	1.23± 0.36[a] 0.22± 0.19[b]	
		Muscle (11)		BDL	BDL	-	BDL	BDL	0.54± 0.19[a] 0.18± 0.10[b]	

Table 1. (Continued)

Species	Locality	Tissue (n)	∑DDT	*o,p'*-DDT	*o,p'*-DDD	*o,p'*-DDE	*p,p'*-DDT	*p,p'*-DDD	*p,p'*-DDE	Source
C. mydas	Baja California peninsula	Liver (7)	BDL- 7.8							Gardner *et al.* (2003)
		Fat (7)	BDL-12.2							
		Kidney (7)	BDL-10.2							
		Muscle (7)	BDL-14.0							
L. olivacea		Liver (1)	10.4							
		Fat (1)	5.1							
		Kidney (1)	18.3							
		Muscle (1)	8.6							
C. caretta		Liver (1)	BDL							
		Fat (1)	BDL							
		Muscle (1)	BDL							
C. caretta	North Carolina	Fat (44)	452 ± 643**	7.03 ± 13.1**				BDL	445 ± 643**	Keller *et al.* (2004a)
		Blood (44)	305 ± 577**	0.28 ± 1.90**				4.1 ± 6.4**	300 ± 578**	
L. kempi		Yellow Fat (10)	156 ± 113**					2.7 ± 4.6**	154 ± 111**	
		Brown Fat (10)	257 ± 332**					2.9 ± 5.1**	254 ± 332**	
		Blood (10)	172 ± 147**					5.9 ± 1.8**	166 ± 147	
C. caretta	Adriatic Sea	Liver (11)		11.6			19.6	16.9	40.6	Perugini *et al.* (2006)
				BDL-21.7			BDL-70.6	BDL-61.2	3.6-217.3	
		Fat (11)		9.5			32.9	26.2	280.9	
				BDL-9.5			BDL-52	BDL-39.6	1.5-621	
		Muscle (11)		0.3			1.3	0.9	4.7	
				BDL-0.5			BDL-2.6	BDL-16	0.4-20.2	
C. caretta	Adriatic Sea	Liver (19)	18.3 ± 18.40							Storelli *et al.* (2007)
		Kidney (19)	5.70 ± 6.49							
		Lung (19)	3.76 ± 7.69							
		Muscle (19)	1.45 ± 1.10							
C. caretta	Canary Islands	Fat (23)	30.7 ± 95.3	23.9 ± 71.2	6.79 ± 25.6	BDL				Monagas *et al.* (2008)
			BDL-449.0	BDL-327.2	BDL-121.8					
		Liver (23)	73.7 ± 174.6	22.2 ± 63.7	48.8 ± 166	2.78 ± 7.7				
			BDL-810.8	BDL-250.9	BDL-798.8	BDL-26.2				

Table 1. (Continued)

Species	Locality	Tissue (n)	∑DDT	*o,p'*-DDT	*o,p'*-DDD	*o,p'*-DDE	*p,p'*-DDT	*p,p'*-DDD	*p,p'*-DDE	Source
L. kempi	Mass.	Plasma (18)					ND->10	ND->10	ND->10	Innis *et al.* (2008)
		Liver (3)					ND	ND-<10	<10-15	
		Kidney (3)					ND	ND	<10	
		Fat (3)					ND	ND	51-209	
		Brain (3)					ND	ND	ND-<10	
L. kempi	Gulf of Mexico	Blood (46)	541 ± 656				33.4± 22.1		472 ± 633	Swarthout *et al.* (2010)
C. mydas		Blood (9)	80 ± 114				56.4± 13.5		66.4 ± 110	
C. caretta	Eastern Adriatic Sea	Fat (27)	112.5 ± 72.3				13 ± 13.9	7 ± 11.1	92.4 ± 53	Lazar *et al.* (2011)
C. mydas	Japan	Liver (5)	<0.02-3.2							Malarvanan *et al.* (2011)
E. imbricata		Liver (3)	10-110							
C. caretta		Liver (4)	15-25							
C. caretta	Florida	Plasma (10)[c]							1,557± 2,656	Ragland *et al.* (2011)
		Plasma (9)[d]							91.7 ± 32.3	
C. mydas	San Diego	Blood (31)							BDL-1.56	Komoroske *et al.* (2011)

BDL: below detection limit.
ND: not detected.
[*] Value expressed as mg/kg lipid weight.
[**] Value expressed as ng/g lipid weight.
[a] Young turtles.
[b] Adult turtles.
[c] Transient turtles.
[d] Resident turtles.

The first paper including data on OC pesticides in organ and tissues of sea turtles was reported by McKim and Johnson (1983). They examined the *p,p'*-DDE levels in the liver and muscle of post yearling green and loggerhead turtles collected along the East coast of Florida. In both sea turtle species, *p,p'*-DDE levels found in the liver were higher than those found in the muscle. In addition, *p,p'*-DDE levels found in loggerheads were higher than those found in green turtles. Authors concluded that OC residues in both species of post yearling sea turtles were present at low concentrations, and differences between species were attributed to the dietary habits of the turtles. The herbivorous green turtle, which feeds primarily on marine sea grasses and algae as an adult, may spend up to the first year of its life as a carnivore, feeding on small molluscs and crustaceans. The loggerhead in comparison is primarily carnivorous feeding on crustacean, conchs, tunicates, and molluscs (McKim and Johnson 1983).

A study focused on Kemp's ridley sea turtle, considered to be the most endangered of all sea turtles, was reported by Lake *et al.* (1994). They measured the concentrations of OC pesticides (DDT compounds, *trans*-nonachlor, and alpha-chlordane) in liver and body fat samples of juvenile Kemp's ridley sea turtles killed by rapid seasonal temperature drops and collected on the eastern shores of Long Island.

In individual samples *trans*-nonachlor concentrations of up to 152 ng/g in liver and from 29 to 332 ng/g in body fat were found. In both the liver and body fat samples, alpha-chlordane was found at low concentrations (<10 ng/g). Because the strong correlations found for concentrations of *p,p'*-DDE and *trans*-nonachlor between the liver and body fat samples, authors concluded that either tissue may be used to monitor these contaminants (Lake *et al.* 1994).

Samples of liver, subcutaneous fat, pectoral muscle, and kidney from juvenile loggerhead and Kemp's ridley sea turtles stranded in Virginia and North Carolina were analyzed to determine concentrations of DDTs (Rybitski *et al.* 1995). Highest concentrations of DDTs were found in the subcutaneous fat and liver, whereas concentrations in the kidney and pectoral muscle were generally below the quantification limit (Rybitski *et al.* 1995).

The first report including data on OC pesticides in sea turtles in Europe was published by Godley *et al.* (1998). They described the mortality patterns of 35 leatherback and three loggerhead turtles entangled in fishing gear, found dead at sea or stranded dead on and around the coast of Britain in the period 1992-1996. However, detection of OC pesticides (DDTs, alpha-HCH, beta-HCH, gamma-HCH, HCB, and dieldrin) was carried out only in the adipose tissue from three leatherback sea turtles. Concentrations of these OC compounds were low. Only *p,p'*-DDE could be detected consistently in all three leatherback turtles (Godley *et al.* 1998).

The first paper including data on OC pesticides in sea turtles from the Mediterranean Sea was reported by Mckenzie *et al.* (1999). They measured the concentrations of OC pesticides [DDTs, endrin, dieldrin, and ∑Chlor (heptachlor + heptachlor epoxide + alpha-chlordane + gamma-chlordane + oxychlordane + *trans*-nonachlor)] in the liver and adipose tissue of nine green sea turtles and five loggerhead turtles from Cyprus and Greece; they also included two leatherback turtles and one loggerhead turtle stranded in Atlantic waters (Scotland). Analysis of DDT and its isomers in the turtle samples showed that *p,p'*-DDE was present in the greatest concentrations, making up >95% of total DDT in the majority of samples. The observed DDT component pattern in most samples was *p,p'*-DDE >*p,p'*-DDT >*p,p'*-DDD, being similar to that observed in most top predators at a distance from DDT point source contamination (Aguilar 1987). The highest *p,p'*-DDE concentrations were detected in loggerhead adipose tissue. The insecticide dieldrin was detected in most of the tissues analyzed, and was often present as the second most abundant OC compound, although at considerably lower concentrations than *p,p'*-DDE. Lindane was detected in two of the loggerhead turtle samples at concentrations < 1ng/g (Mckenzie *et al.* 1999).

Samples of liver, lung, kidney, and muscle from 11 loggerhead sea turtles stranded along the Adriatic Sea (Italy) were analyzed to determine concentrations of OC pollutants (DDTs and HCB) (Storelli and Marcotrigiano 2000). Among DDT metabolites only *p,p'*-DDE was found in this study. DDT is mostly metabolized by the action of microsomal enzymes. DDT is transformed into DDE through a dehydrochlorination produced by decomposition reactions, and consequently DDE becomes progressively more abundant; more, DDE on account of its low polarity is excreted with difficulty and it inclines to accumulate (Storelli and Marcotrigiano 2000). In addition, the highest *p,p'*-DDE values were detected in juveniles with respect to adult turtles. The concentrations of HCB were the lowest of all OC pollutants quantified in all tissues. Authors concluded that the significant differences between the levels of OC compounds in tissues from juveniles and adult females could suggest that exists a mobilization of OC compounds through the excretion of these contaminants via reproductive activities (Storelli and Marcotrigiano 2000).

The first paper including data on OC pesticides in sea turtles from the Eastern Pacific was reported by Gardner *et al.* (2003). They measured the concentrations of OC pesticides [DDTs, HCB, aldrin/dieldrin, endosulfan I and II, endrin, lindane, ∑Chlor (heptachlor + heptachlor epoxide + *cis*-chlordane + *trans*-chlordane + oxychlordane + *cis*-nonachlor + *trans*-nonachlor)] in the liver, kidney, muscle, and adipose tissue of seven green, one loggerhead, and one olive ridley sea turtles that died as a result of incidental fisheries capture in Baja California, Mexico. The sea turtle tissues analyzed in this study contained low levels of OC pesticides, and the authors suggested that the no-industrialization of the region could explain the results. Heptachlor epoxide was the most prevalent of 41 OC compounds examined in this study and was detected in 40% of the tissue samples. In addition, the pattern of accumulation of the isomers of DDT was different: in this study *p,p'*-DDD contributed most to ∑DDT, unlike previous reports for sea turtles collected elsewhere, wherein *p,p'*-DDE was the dominant of the DDT isomers.

The first papers reporting concentrations of OC pesticides in sea turtle blood were published by Keller at al. (2004a,b). Firstly, they compared three different extraction techniques to determine the relative efficiency of extracting OC contaminants from loggerhead sea turtle whole blood. Extraction of OC pesticides (DDTs, mirex, dieldrin, oxychlordane, *trans*-nonachlor, total chlordanes) from whole plasma, plasma, and red blood cell samples from five loggerhead sea turtles captured as bycatch in North Carolina was conducted. Plasma contained the highest OC pesticide concentrations, followed by whole blood and red blood cells. The results of this interesting study indicated that blood, preferably plasma, could be used to detect and monitor OC contaminants in loggerhead sea turtles (Keller *et al.* 2004b). In addition, to determine whether blood OC concentrations can predict those in adipose tissue of sea turtles, paired fat and blood samples were collected from 44 live, juvenile loggerhead sea turtles and 10 juvenile Kemp's ridley sea turtle carcasses. Lipid-normalized OC concentrations measured in the blood significantly correlated to levels found in the fat samples of both species. However, blood OC concentrations calculated on a wet-mass basis were significantly and inversely correlated to lipid content in the fat samples. This is important, because as authors remarked, caution should be used when monitoring spatial or temporal trends, as OC levels may increase in the blood following mobilization of fat stores, such as during long migrations, breeding, or diseases (Keller *et al.* 2004a).

Samples of liver, muscle, and fat from 11 loggerhead sea turtles stranded in the central and southern cost of the Adriatic Sea (Italy) were analyzed to determine concentrations of OC pesticides (DDTs, alpha-endosulfan, beta-endosulfan, alpha-HCH, beta-HCH, gamma-HCH, heptachlor, heptachlor epoxide, HCB, and methoxychlor) (Perugini *et al.* 2006). Only four were detected: *p,p'*-DDE (57%), *p,p'*-DDD (16%), and *p,p'*-DDT and *o,p'*-DDT (27%). The order of concentrations of pesticides was: *p,p'*-DDE >*p,p'*-DDT >*p,p'*-DDD >*o,p'*-DDT in all tissues. The contamination levels of OC pesticides among tissues showed the following pattern: fat > liver > muscle. In this study *p,p'*-DDE represented 46% (in liver), 80% (in fat), and 64% (in muscle) of the total DDTs (Perugini *et al.* 2006).

Samples of liver, kidney, lung, and muscle tissue from 19 loggerhead turtles stranded in different coastal areas of the Adriatic and Ionian Sea were analyzed to determine concentrations of DDT and isomers (Storelli *et al.* 2007). The highest levels of these compounds were found in liver, followed by kidney, lung, and muscle tissue, likely reflecting the relative lipid concentrations of these tissues. Among DDTs, the compound in the greatest concentration was *p,p'*-DDE (liver: 85.2%, kidney: 93.6%, lung: 86.4%, muscle tissue: 93.2%) (Storelli *et al.* 2007).

The first paper reporting OC pesticide levels in sea turtles in West Africa was published by Monagas *et al.* (2008). They measured the concentrations of *o,p'*-DDT, *o,p'*-DDD, and *o,p'*-DDE in the liver and adipose tissue of 23 loggerhead sea turtles stranded on the coasts of the Canary Islands (Spain). The pattern of distribution was different among liver and fat. In the liver the most abundant was *o,p'*-DDD (66.1%), followed by *o,p'*-DDT (30%), and *o,p'*-DDE (3.7%). In the fat the most abundant was *o,p'*-DDT (78%), followed by *o,p'*-DDD (22%). Because the majority of previous surveys have been dedicated to *p,p'*-DDT compounds, this paper reported important information on o,p'-DDT compounds. Authors attributed the predominance of *o,p'*-DDD to a species-specific metabolic capability and/or specific exposure in reductive (anoxic) conditions (Monagas *et al.* 2008). The very low concentration for *o,p'*-DDE had been reported previously in human beings from the Canary Islands (Zumbado *et al.* 2005). In addition, Monagas *et al.* (2008) reported in the liver ∑DDT concentrations higher than in the adipose tissue. Eleven turtles showed ∑DDT levels in the liver higher than in the fat; all these turtles showed cachexia. Authors suggested that the cachexia observed in the turtles could indicate a previous remobilization of the fat and the pollutants, explaining the high ∑DDT in the liver (Monagas *et al.* 2008).

A new study focused on cold-stunned juvenile Kemp's ridley sea turtles from Cape Cod (Massachusetts) was reported by Innis *et al.* (2008). They measured the concentrations of 18 OC pesticides in plasma of 18 live turtles, and liver, kidney, fat, and brain of three dead turtles. Aldrin, alpha-HCH, alpha-chlordane, beta-HCH, delta-HCH, endosulfan I, endosulfan II, endosulfan sulphate, endrin, endrin aldehyde, endrin ketone, heptachlor, heptachlor epoxide, and methoxychlor were not detected in any sample. Dieldrin, gamma-HCH, gamma-chlordane, *p,p'*-DDD, *p,p'*-DDE, and *p,p'*-DDT were detected at low concentrations in few turtles (Innis *et al.* 2008).

Blood samples from 46 juvenile and sub-adult Kemp's ridley and 9 green sea turtles from the Gulf of Mexico were analyzed to detect 25 OC pesticides (Swarthout *et al.* 2010). ∑DDTs and ∑chlordanes were significantly higher in Kemp's ridley than in green turtles, likely due to differences in dietary preferences and biomagnifications potential: green turtles are primarily herbivorous whereas Kemp's ridleys primarily consume benthic invertebrates. Concentrations of *p,p'*-DDT were detected in 44% of the green turtles and 59% of the Kemp's ridley turtles. The mean ratio of *p,p'*-DDT to *p,p'*-DDE concentrations was higher in green turtles than in Kemp's ridleys. Authors suggested that differences in the ability to metabolize the parent compound *p,p'*-DDT to *p,p'*-DDE could explain this fact (Swarthout *et al.* 2010).

Samples of fat tissue from 27 loggerhead sea turtles stranded along the eastern Adriatic Sea (Croatia) were analyzed to determine concentrations of OC pollutants (DDTs, HCB, and alpha-HCH) (Lazar *et al.* 2011). Among DDTs, *p,p'*-DDE accounted for 80% of the total DDTs. HCB was measurable in all samples, with a median concentration of about two fold higher than in the western Atlantic Ocean. In addition, alpha-HCH was present in 89% of samples (Lazar *et al.* 2011).

The first paper including data on OC pesticides in sea turtles from Japan was recently reported by Malarvannan *et al.* (2011).They measured the concentrations of OC pesticides [DDTs, HCHs, HCB, and CHLs (oxychlordane, *trans*-nonachlor, and *cis*-nonachlor)] in the liver of five green, four loggerhead, and three hawksbill sea turtles stranded along the coasts or caught (by-catch) around Ishigaki Island and Kochi. Among OC pesticides, DDTs and CHLs were the predominantly compounds in all the three turtle species. Among DDTs, *p,p'*-DDE was the predominant compound, suggesting a wide use and long-term accumulation of DDTs in all the three species in Japan. Mean concentrations of HCHs and HCB compounds were several orders of magnitude lower than those of DDTs. The concentrations of OC contaminants were lower than concentrations reported in sea turtles collected in other locations around the world. Authors suggested that samples were obtained from an unindustrialized region (Malarvannan *et al.* 2011).

Recently, OC pesticides (*p,p'*-DDE, mirex, *trans*-chlordane, *trans*-nonachlor, oxychlordane, and toxaphenes) were measured in blood plasma of satellite-tracked adult male loggerhead sea turtles captured in Cape Canaveral, Florida (Ragland *et al.* 2011). Nine turtles were resident, whereas 10 transient turtles migrated northward. Transients showed elevated blood plasma concentrations of numerous organic pollutants compared with residents. Detection frequencies in the transient group were equal to or greater than those in the resident group for all organic pollutants, except for *p,p'*-DDE. Turtles from the transient group showed significantly elevated concentrations of mirex, *trans*-chlordane, *trans*-nonachlor, and oxychlordane compared with turtles from the resident group. Authors suggested that foraging location could influence exposure to, and patterns of, organic pollutants in highly mobile species such as loggerheads (Ragland *et al.* 2011).

Also recently, OC pesticides (*p,p'*-DDE, gamma-HCH, heptachlor epoxide, alpha-chlordane, gamma-chlordane) were measured in blood plasma of 31 green sea turtles resident in an urbanized estuary in San Diego, California (Komoroske *et al.* 2011). Concentrations of *p,p'*-DDE and gamma-chlordane were higher relative to other sea turtle studies.

Table 2. Pesticide concentrations (average concentration or range, ng/g w. w.) in sea turtle eggs

Species/Locality	Eggs (n)[a]	*p,p'*-DDE	∑DDTs	∑chlor	Mirex	Dieldrin	∑HCH	Source
C. caretta/ S.Carolina and Ga.	NR		58-305			ND-56.4		Hillestad *et al.* (1974)
C. mydas/ Ascension Island	10 (4)	3±1 ND-9.0						Thompson *et al.* (1974)
C. caretta/ Florida	(9)	66±21 18-200		ND-6,17, or 9[b]	ND-5	ND-28		Clark and Krynitsky (1980)
C. mydas/ Florida	(2)	2±2 ND-5		ND	ND	ND		
C. caretta/ Florida	56 (1)	99 56-150		ND-8				Clark and Krynitsky (1985)
C. mydas/ Australia	15 (4)	1.7 1.3-2.4						
C. caretta/ Cyprus	1		155	1.8		0.6		Mackenzie *et al.* (1999)
C. mydas/ Cyprus	1		4.3	<0.3		<0.3		
C. caretta/ Florida	4-12 (20)		ND-178[c]	ND	ND	ND	ND	Alam and Brim (2000)
C. caretta/ Florida	22 (3)		50.2±92 1.19-344	25.5±46.7 0.827-175	1.03±2.04 0.009-7.52	2.53±2.76 0.278-10.8	0.258±0.5 0.02-2.05	Alava *et al.* (2006)
C. mydas/ Malaysia	55 (NR)		0.08±0.01 BDL-0.7	0.057±0.1 0.02-0.51			0.069 0.01-0.23	Van de Merwe *et al.* (2009)
C. mydas/ Malaysia	NR (11)			0.018[d] 0.01-0.02	0.009 BDL-0.012		0.17[e] 0.13-0.20	Van de Merwe *et al.* (2010)
C. caretta/ W. Florida	NR (11)	22.7±7 0.81-74	23.8±7.1 2.36-74	20.8±9.6 <0.4-106	1.04±0.53 <0.09-5.61	5.06±1.09 1.79-14.7	0.45 <0.4-1.09	Alava *et al.* (2011)
E. Florida	NR (23)	135±56 0.7-1030	136±56 0.78-1030	113±31 0.73-558	6.78±3.78 <0.09-90.2	10±1.9 <1.14-32	1.21±0.49 <0.4-10.4	
N. Carolina	NR (9)	690±250 1.8-2170	694±251 4.97-2170	375±146 3.85-1280	10.3±3 0.45-29.7	29.9±11 1.98-76.1	3.15±1.39 <0.5-13.1	

[a] n= total number of eggs analyzed individually from a total number of nests in parentheses.
[b] Range of heptachlor epoxide, oxychlordane, and *trans*-nonachlor, respectively.
[c] Reported as *p,p'*-DDE alone.
[d] Reported as *trans*-chlordane alone.
[e] Reported as *gamma*-HCH alone.
ND: not detected.
NR: not reported.

ORGANOCHLORINE PESTICIDES IN SEA TURTLE EGGS

The first studies on OC pesticides in sea turtles were focused on the analysis of OC contaminants in eggs (Table 2), as previously had been studied in bird species. The first surveys were published in 1974 (Hillestad *et al.* 1974, Thompson *et al.* 1974). Whereas eggs of green turtles from Ascension Island contained only very low OC residues (Thompson *et al.* 1974), residues in eggs of loggerhead turtles from Georgia and South Carolina appeared to be somewhat higher (Hillestad *et al.* 1974). In other study, eggs from nine clutches of loggerhead turtles and two clutches of green turtles were collected in Florida, incubated, and analyzed for OC pesticides (DDTs, dieldrin, endrin, heptachlor epoxide, mirex, oxychlordane, *cis*-chlordane, *trans*-nonachlor, *cis*-nonachlor, HCB, and toxaphene) (Clark and Krynitsky 1980). Residues of eight OC compounds were recovered from eggs of loggerhead turtles, but only DDE and DDT were found in eggs of green turtles. Residue concentrations were remarkably low, probably reflecting low dietary intake (Clark and Krynitsky 1980). The same authors reported higher DDE concentrations in other study (far below levels thought to be harmful), also demonstrating that DDE level did not decline during incubation (Clark and Krynitsky 1985).

In the study on concentrations and patterns of OC contaminants in sea turtles from Mediterranean and Atlantic waters (Mckenzie *et al.* 1999), a small number of eggs were collected from nests hatched at northern Cyprus. Authors concluded that loggerhead turtles were exposed to relatively higher concentrations of OC pesticides, particularly *p,p'*-DDE, during early developmental stages (Mckenzie *et al.* 1999).

Alam and Brim (2000) analyzed unhatched loggerhead turtle eggs from 20 nests along northwest Florida for 21 OC pesticides, but only *p,p'*-DDD was detected. It was present at two nesting sites, at concentrations of 800 and 753 ng/g, respectively, concluding that OC pesticides alone did not appear to be a limiting factor for sea turtles eggs hatching within the study area (Alam and Brim 2000).

OC pesticides were measured in yolk samples from eggs that failed to hatch from three loggerhead sea turtle nests collected in southern Florida to determine change of concentrations through embryonic development (Alava *et al.* 2006). Mean concentrations of ∑DDTs, ∑chlordanes, and dieldrin in all eggs were 67.1, 37.0, 11.1 ng/g lipid, respectively. Early and middle developmental stage samples had similar concentrations of OC pesticides on a wet-mass basis (ng/g tissue extracted), but the concentrations doubled by the late stage. Authors concluded that this increase was mort likely attributable to the 50% increase in lipid content observed in the late-stage yolk, indicating that an early-stage sample cannot be directly compared to a late-stage sample, especially from different nests (Alava *et al.* 2006).

Green turtle eggs were collected from markets in Peninsular Malaysia and analyzed for 23 OC pesticides (van de Merwe *et al.* 2009). Compounds most commonly identified were dieldrin, endosulfan I, heptachlor epoxide, HCB, mirex, beta-HCH, gamma-HCH, *p,p'*-DDE, and *p,p'*-DDT. The concentrations of OC pesticides found in this study were lower than concentrations previously reported. Although hazard quotients for each compound indicated minimal risk for the embryos, authors suggested that the large number of OC compounds detected in the eggs could be important in terms of combined effects (van de Merwe *et al.* 2009). More recently, in a study on maternal transfer, eggs from 11 green turtles from Malaysia were analyzed for 23 OC pesticides (van de Merwe *et al.* 2010). The chemical profiles in eggs from individual turtles were significantly different, indicating variable contaminant uptake during foraging. The only OC pesticides detected were *trans*-chlordane, mirex, and gamma-HCH. The authors concluded that the most alarming statistic from the assessment of the human health risk associated with consuming green turtle eggs in Peninsular Malaysia was related to coplanar PCBs, not OC pesticides (van de Merwe *et al.* 2010).

Recently, OC pesticides (DDTs, chlordanes, mirex, dieldrin, HCHs, HCB, and toxaphene congeners) were measured in loggerhead turtle egg yolk samples collected from 44 nests in three distinct U. S. locations: North Carolina, eastern Florida, and western Florida (Alava *et al.* 2011). Mirex was significantly higher in concentration in North Carolina than western Florida, with eastern Florida being intermediate. The predominant chlordanes (*trans*-nonachlor, oxychlordane, and heptachlor epoxide) resulted in total chlordanes being significantly lower in western Florida than in the other regions. The predominant DDT metabolite in all samples was the *p,p'*-DDE, and its concentrations were significantly higher in North Carolina than western Florida, with eastern Florida having intermediate concentrations (Alava *et al.* 2011).

PESTICIDES AND SEXUAL DIFFERENTIATION IN SEA TURTLES

In many turtles, the temperature experienced during the middle third of incubation determines the sex of the offspring (Ewert and Nelson 1991). The implication of steroid sex hormones as the proximate trigger for sex determination opens the possibility that endocrine-disrupting contaminants could also influence the outcome of sexual differentiation. An interesting study on the potential effects of DDE on sexual differentiation of green sea turtles was reported by Podreka *et al.* (1998). Four clutches of eggs collected from Queensland, Australia, were treated with *p,p'*-DDE at the beginning of the thermosensitive period for sexual determination. Dosed eggs were incubated at two temperatures on the upper and lower boundaries of the sex determination threshold for this species. Treated eggs, corrected for background *p,p'*-DDE, had up to 543 ng/g *p,p'*-DDE. The sex ratio at these doses did not differ from what would be expected on consideration of temperature alone. Incubation time, hatching success, incidence of body deformities, hatching size, and weight were also within the limits of healthy developed hatchlings. Authors demonstrated that the eggs of green turtles in the wild with *p,p'*-DDE concentrations < 543 ng/g should produce hatchlings with relatively high hatching success, survival rate, and normally differentiated gonads (Podreka *et al.* 1998).

In an *in vitro* study on the effects of pesticides (atrazine, *p,p'*-DDE) on cytochrome P450 aromatase, the key enzyme responsible for the conversion of testosterone to 17beta-estradiol, in an immortal testis cell line from a green sea turtle, atrazine significantly induced aromatase activity following a 24 h exposure, and p,p'-DDE inhibited the activity but only at cytotoxic concentrations (Keller and McClellan-Green 2004).

More recently, the effect of OC pesticides (DDT, DDE, and dieldrin) on sex steroid-binding proteins in vitro in the plasma of nesting green turtles from Peninsular Malaysia was reported (Ikonomopoulou *et al.* 2009). Authors demonstrated that dieldrin competed *in vitro* with estradiol for binding sites. Testosterone binding affinity was affected at various DDT concentrations and was hypothesized that DDT *in vivo* could act to inhibit steroid-protein interactions in nesting green turtles. Authors suggested that DDT could have an effect upon the protein conformation thus affecting T binding (e.g. the T binding site on the steroid hormone binding protein molecule) (Ikonomopoulou *et al.* 2009).

MATERNAL TRANSFER OF PESTICIDES IN SEA TURTLES

The maternal transfer of five OC pesticides (*p,p'*-DDT, *p,p'*-DDE, *p,p'*-DDD, alpha-HCH, and gamma-HCH), and the temporal variation of blood and eggs concentrations from 38 leatherback sea turtles nesting in French Guiana, was recently reported by Guirlet *et al.* (2010). Leatherbacks in this study presented low OC pesticide concentrations in blood and eggs probably resulting from their low trophic level with a diet based mainly on jellyfish, and from the pollution status of their foraging and nesting grounds. All OC pesticides detected in leatherback blood were detected in eggs, suggesting a maternal transfer of OC pesticides. This transfer was shown to depend on female blood concentration for ∑DDTs, since significant correlations were found between paired blood-egg concentrations. During the nesting season, OC pesticides concentrations in eggs and the percentage of lipid in eggs were found to decline in successive clutches, highlighting a process of offloading from females to their eggs and a decreasing investment of lipid from females into their clutches (Guirlet *et al.* 2010).

The maternal transfer of OC pesticides was also investigated in green turtles from Malaysia by van de Merwe *et al.* (2010). Authors demonstrated maternal transfer of OC pesticides to eggs and hatchlings, with significant correlations in gamma-HCH, *trans*-chlordane and mirex concentrations between maternal blood and eggs, between eggs and hatchling blood, and between maternal and hatchling blood. In addition, there was also a significant correlation between increasing egg persistent organic pollutants (including OC pesticides, PCBs, and PBDEs) concentration and decreasing hatchling mass/length ratio (van de Merwe *et al.* 2010).

EFFECTS OF ORGANOCHLORINE PESTICIDES ON SEA TURTLE IMMUNITY

Keller *et al.* (2006) demonstrated that certain lymphocyte proliferation responses in loggerhead sea turtles were positively correlated with OC pesticide concentrations measured in blood, even though the concentrations in sea turtles were generally much lower than in fish-eating wildlife. Lysozyme activity was significantly and negatively correlated with whole blood concentrations of *p,p'*-DDE and the sum of chlordanes. Mitogen-induced lymphocyte proliferations were also significantly and positively correlated with *p,p'*-DDE blood concentrations. Authors concluded that these correlative observations in free-ranging turtles suggest that chronic exposure to OC pesticides could suppress innate immunity and enhance certain lymphocyte functions of loggerhead sea turtles. To test this hypothesis, authors measured lymphocyte proliferation after *in vitro* exposure of peripheral blood leukocytes from 16 turtles to *p,p'*-DDE. This contaminant increased mitogen-induced proliferation at concentrations below those than affected cell viability. These results were similar to the findings of many other wildlife studies and suggest that the sea turtle immune system is modulated by environmentally relevant concentrations of OC pesticides (Keller *et al.* 2006).

More recently, in the study on pollutants and the health of green sea turtles resident to an urbanized estuary in San Diego (Komoroske *et al.* 2011), authors found that blood *p,p'*-DDE concentrations exceeded lymphocyte proliferation no-effects levels, suggesting that immunological effects could be a concern at current concentrations of this pollutant.

OC Pesticides and Clinical Health Parameters and/or Diseases

The first study conducted to determine the possible involvement of pesticides in a sea turtle disease was reported by Aguirre *et al.* (1994). Because environmental contaminants have been listed as a possible cause of green turtle fibropapillomatosis, authors tested samples of brain, fat, liver, and kidney tissues from 10 juvenile green turtles afflicted with fibropapillomatosis to determine 23 OC pesticides and other environmental contaminants. In addition, they also tested egg shells and tissues from three green turtle hatchlings. The tissues and shells analyzed in this study indicated that none contained any of the listed OC pesticides in concentrations above the stated method of detection limits. No correlation was found between the contaminants tested and fibropapillomatosis because the low levels detected. Authors concluded that environmental contaminants did not appear to play a role in the etiology of green turtle fibropapillomatosis (Aguirre *et al.* 1994).

OC pesticide concentrations in loggerhead sea turtles were compared with clinical health parameters (hematology, plasma chemistry, and body condition) (Keller *et al.* 2004c). Concentrations of ∑DDTs, ∑chlordanes, dieldrin, and mirex were determined in 44 fat biopsies and 48 blood samples. Blood concentrations of ∑chlordanes were negatively correlated with red blood cell counts, hemoglobin, and hematocrit, indicative of anemia. Positive correlations were observed between most classes of OC pesticides and white blood cell counts and between mirex and the heterophil/lymphocyte ratio, suggesting modulation of the immune system. All classes of OC pesticides in the blood except dieldrin were correlated positively with aspartate aminotransferase activity, indicating possible hepatocellular damage. Mirex blood concentration was negatively correlated with alkaline phosphatase activity. Glucose concentrations were negatively correlated with adipose concentrations of dieldrin and ∑DDTs. The ratio albumin/globulin was negatively correlated with concentrations of ∑chlordanes in blood and fat. Blood urea nitrogen (BUN) concentrations were positively correlated with concentrations of most OC pesticides measured in blood. Sodium concentrations were positively correlated with blood concentrations of mirex. Magnesium concentrations were negatively correlated with ∑DDTs in the blood. Body condition was negatively correlated with dieldrin in the blood. Authors concluded that OC pesticides could affect the health of loggerhead turtles even though sea turtles accumulate lower concentrations of OC pesticides compared with other wildlife (Keller *et al.* 2004c).

In the survey of OC pesticide baseline levels in loggerhead sea turtles in the Canary Islands (Monagas *et al.* 2008), authors also studied the correlation with body condition and tissue lesions. Emaciated turtles showed higher average ∑DDT concentrations in the liver than in the fat. Histologically, they detected hepatic lesions in four loggerhead turtles with high ∑DDT concentrations (severe multifocal necrotic hepatitis and diffuse vacuolar hepatic degeneration), but these were not exclusively attributed to acute effects of DDTs because clinical history of these turtles also included severe septicemia. However, authors did not discard chronic effects of DDTs (Monagas *et al.* 2008).

In the survey of OC pesticides in blood of Kemp's ridley and green sea turtles from the Gulf of Mexico (Swarthout *et al.* 2010), they found some correlations with morphological, health and immune function parameters. Significant negative correlations were observed between ∑DDTs and the proliferative response of T-cells to phytohemagglutinin P, and between dieldrin and testosterone concentrations in female turtles. BUN concentrations were positively correlated with ∑chlordane concentrations, suggesting that increased BUN in turtles is less of an indicator of renal disease and more of an indicator of recent feeding (Swarthout *et al.* 2010).

CONCLUSION

Among the studies on baseline levels of OC pesticides in sea turtles, the majority of surveys reported values for DDTs in organs and tissues. Liver and fat are the matrices of choice for the investigation of OC pesticides, although blood samples, preferably plasma, have been used since 2004 because blood can be collected easily from free-ranging populations, facilitating the repeated collection of larger number of samples. The most frequent pattern of distribution of DDTs reported in sea turtles is: adipose tissue > liver > muscle. However, cachectic turtles can have higher concentrations of DDTs in the liver because the remobilization of the fat and the pollutants. Concentrations of OC pesticides in blood may better represent the exposure levels of target tissues, but blood concentrations may fluctuate more than those in adipose tissues following recent dietary exposure or lipid mobilization.

The scarce studies to determine baseline levels of OC pesticides in sea turtles eggs showed low concentrations of OC pesticides, although a large number of OC compounds detected in the eggs could be important in terms of combined effects. In addition, more research on the effects of pesticides on sexual differentiation of sea turtles is necessary.

Maternal transfer of OC pesticides was demonstrated in leatherback and green sea turtles. This transfer was shown to depend on female blood concentration for OC pesticides, since significant correlations were found between paired blood-egg concentrations.

The scarce studies to determine the effects of OC pesticides on the immune system of sea turtles resulted in very similar findings to those reported in other wildlife studies, suggesting that the sea turtle immune system is modulated by environmentally relevant concentrations of OC pesticides.

Currently, a correlation between OC pesticides and health status of sea turtles is difficult to establish because of scarce number of studies, and the numerous factors affecting the sea turtles health. Nevertheless, clinical health parameters (hematology, plasma chemistry, and body condition) in loggerheads were affected with OC pesticide concentrations.

Continued monitoring of pesticides coupled with immunological, physiological, and pathological effects studies will increase our understanding of the impact of these pollutants on endangered sea turtles.

REFERENCES

Aguilar, A. (1987). Using organochlorine pollutants to discriminate marine mammal populations: a review and critique of methods. *Marine Mammal Science*, 3, 242-262.

Aguirre, A. A., Balazs, G. H., Zimmerman, B. and Galey, F. D. (1994). Organic contaminants and trace metals in the tissues of green turtles (*Chelonia mydas*) afflicted with fibropapillomas in the Hawaiian Islands. *Marine Pollution Bulletin*, 28, 109-114.

Alam, S. K. and Brim, M. S. (2000). Organochlorine, PCB, PAH, and metal concentrations in eggs of loggerhead sea turtles (*Caretta caretta*) from Northwest Florida, USA. *Journal of Environmental Science and Health*, Part B, 35, 705-724.

Álava, J. J., Keller, J. M., Kucklick, J. R., Wyneken, J., Crowder, L. and Scott, G. I. (2006). Loggerhead sea turtle (*Caretta caretta*) egg yolk concentrations of persistent organic pollutants and lipid increase during the last stage of embryonic development. *Science of the Total Environment*, 367, 170-181.

Álava, J. J., Keller, J. M., Wyneken, J., Crowder, L., Scott, G. and Kucklick, J. R. (2011). Geographical variation of persistent organic pollutants in eggs of threatened loggerhead sea turtles (*Caretta caretta*) from Southeastern United States. *Environmental Toxicology and Chemistry*, 30, 1677-1688.

Bolt, H. M. and Degen, G. H. (2002). Comparative assessment of endocrine modulators with estrogenic activity: II. Persistent organochlorine pollutants. *Archives of Toxicology*, 76, 187-193.

Clark, D. R. Jr. and Krynitsky, A. J. (1980). Organochlorine residues in eggs of loggerhead and green sea turtles nesting at Merritt Island, Florida-July and August 1976. *Pesticides Monitoring Journal*, 14, 7-10.

Clark, D. R. Jr. and Krynitsky, A. J. (1985). DDE residues and artificial incubation of loggerhead sea turtles. *Bulletin of Environmental Contamination and Toxicology*, 34, 121-125.

D'Ilio, S., Mattei, D., Blasi, M. F., Alimonti, A. and Bogialli, S. (2011). The occurrence of chemical elements and POPs in loggerhead turtles (*Caretta caretta*): An overview. *Marine Pollution Bulletin*, 62, 1606-1615.

Ecobichon, D. J. (1995). Toxic effects of pesticides: organochlorine insecticides. In C. D. Klaassen, M. O. Amdur and J. Doull (Eds.), Casarett and Doull's Toxicology: *The basic science of poisons* (pp. 649-655). New York, NY: McGraw-Hill.

Ewert, M. A. and Nelson, C. E. (1991). Sex determination in turtles: diverse patterns and some possible adaptive values. *Copeia*, 1991, 50-69.

Gardner, S. C., Pier, M. D., Wesselman, R. and Juárez, J. A. (2003). Organochlorine contaminants in sea turtles from the Eastern Pacific. *Marine Pollution Bulletin*, 46, 1082-1089.

Godley, B. J., Gaywood, M. J., Law, R. J., McCarthy, C. J., McKenzie, C., Patterson, I. A. P., Penrose, R. S., Reid, R. J. and Ross, H. M. (1998). Patterns of marine turtles mortality in

British waters (1992-1996) with reference to tissue contaminant levels. *Journal of the Marine Biological Association of the United Kingdom*, 78, 973-984.

Guirlet, E., Das, K, Thomé, J. P. and Girondot, M. (2010). Maternal transfer of chlorinated contaminants in the leatherback turtles, *Dermochelys coriacea*, nesting in French Guiana. *Chemosphere*, 79, 720-726.

Hillestad, H. O., Reimold, R. J., Stickney, R. R., Windom, H. L. and Jenkins, J. H. (1974).Pesticides, heavy metals, and radionuclide uptake in loggerhead sea turtles from South Carolina and Georgia. *Herpetological* Review, 5, 75.

Ikonomopoulou, M. P., Olszowy, H., Hodge, H. and Bradley, A. J. (2009). The effect of organochlorine and heavy metals on sex steroid-binding proteins in vitro in the plasma of nesting green turtles, *Chelonia mydas*. *Journal of Comparative Physiology B*, 179, 653-662.

Innis, C., Tlusty, M., Perkins, C., Holladay, S., Merigo, C. and Weber III, E. S. (2008). Trace metal and organochlorine pesticide concentrations in cold-stunned juvenile Kemp's ridley turtles (*Lepidochelys kempii*) from Cape Cod, Massachussets. *Chelonian Conservation and Biology*, 7, 230-239.

IUCN/SSC (1995). A Global Strategy for the Conservation of Marine Turtles. Balmar, Arlington, VA: International Union for Conservation of Nature and Natural Resources.

IUCN/SSC (2012). 2011 IUCN Red List of Threatened Species. *www.iucnredlist.org*. Accessed January 4, 2012.

Iwata, H., Tanabe, S., Sakai, N., Nishimura, A. and Tatsukawa, R. (1994). Geographical distribution of persistent organochlorines in air, water and sediments from Asia and Oceania, and their implications for global redistribution from lower latitudes. *Environmental Pollution*, 85, 15-33.

Keller, J. M., Kucklick, J. R., Harms, C. A. and McClellan-Green, P. D. (2004a). Organochlorine contaminants in sea turtles: correlations between whole blood and fat. *Environmental Toxicology and Chemistry*, 23, 726-738.

Keller, J. M., Kucklick, J. R. and McClellan-Green, P. D. (2004b). Organochlorine contaminants in loggerhead sea turtle blood:extraction techniques and distribution among plasma and red blood cells. *Archives of Environmental Contamination and Toxicology*, 46, 254-264.

Keller, J. M., Kucklick, J. R., Stamper, M. A., Harms, C. A. and McClellan-Green, P. D. (2004c). Associations between organochlorine contaminant concentrations and clinical health parameters in loggerhead sea turtles from North Carolina, USA. *Environmental Health Perspectives*, 112, 1074-1079.

Keller, J. M. and McClellan-Green, P. (2004). Effects of organochlorine compounds on cytochrome P450 aromatase activity in an inmortal sea turtle cell line. *Marine Environmental Research*, 58, 347-351.

Keller, J. M., McClellan-Green, P. D., Kucklick, J. R., Keil, D. E. and Peden-Adams, M. M. (2006). Effects of organochlorine contaminants on loggerhead sea turtle immunity: comparison of a correlative field study and in vitro exposure experiments. *Environmental Health Perspectives,* 114, 70-76.

Komoroske, L. M., Lewison, R. L., Seminoff, J. A., Deheyn, D. D. and Dutton, P. H. (2011). Pollutants and the health of green sea turtles resident to an urbanized estuary in San Diego, CA. *Chemosphere,* 84, 544-552.

Lake, J. L., Haebler, R., McKinney, R., Lake, C. A. and Sadove, S. S. (1994). PCBs and other chlorinated organic contaminants in tissues of juvenile Kemp's ridley turtles (*Lepidochelys kempi*). *Marine Environmental Research*, 28, 313-327.

Lazar, B., Maslov, L., Romanic, S. H., Gracan, R., Krauthacker, B., Holcer, D. and Tvrtkovic, N. (2011). Accumulation of organochlorine contaminants in loggerhead sea turtles, *Caretta caretta*, from the Eastern Adriatic Sea. *Chemosphere*, 82, 121-129.

Malarvannan, G., Takahashi, S., Isobe, T., Kunisue, T., Sudaryanto, A., Miyagi, T., Nakamura, M., Yasumura, S. and Tanabe, S. (2011). Levels and distribution of polybrominated diphenyl ethers and organochlorine compounds in sea turtles from Japan. *Marine Pollution Bulletin*, 63, 172-178.

Mckenzie, C., Godley, B. J., Furness, R. W. and Wells, D. E. (1999). Concentrations and patterns of organochlorine contaminants in marine turtles from Mediterranean and Atlantic waters. *Marine Environmental Research,* 47, 117-135.

McKim, J. M. Jr. and Johnson, K. L. (1983). Polychlorinated biphenyls and *p,p'*-DDE in loggerhead and green postyearling Atlantic sea turtles. *Bulletin of Environmental Contamination and Toxicology*, 31, 53-60.

Monagas, P., Orós, J., Araña, J. and González-Díaz, O. M. (2008). Organochlorine pesticide levels in loggerhead turtles (*Caretta caretta*) stranded in the Canary Islands, Spain. *Marine Pollution Bulletin*, 56, 1949-1956.

Perugini, M., Giammarino, A., Olivieri, V., Guccione, S., Lai, O. R. and Amorena, M. (2006). Polychlorinated biphenyls and organochlorine pesticide levels in tissues of *Caretta caretta* from the Adriatic Sea. *Diseases of Aquatic Organisms*, 71, 155-161.

Podreka, S., Georges, A., Maher, B. and Limpus, C. J. (1998). The environmental contaminant DDE fails to influence the outcome of sexual differentiation in the marine turtle *Chelonia mydas*. *Environmental Health Perspectives,* 106, 185-188.

Pritchard, P. C. H. (1997). Evolution, phylogeny, and current status. In P. L. Lutz and Musick, J. A. (Eds.), *The Biology of Sea Turtles* (pp. 1-28). Boca Raton, FL: CRC Press.

Pugh, R. S. and Becker, P. R. (2001). Sea Turtle Contaminants: a Review with Annotated Bibliography. Report NISTIR 6700. Charleston, SC: National Institute of Standards and Technology.

Ragland, J. M., Arendt, M. D., Kucklick, J. R. and Keller, J. M. (2011). Persistent organic pollutants in blood plasma of satellite-tracked adult male loggerhead sea turtle (*Caretta caretta*). *Environmental Toxicology and Chemistry,* 30, 1549-1556.

Rybitski, M. J., Hale, R. C. and Musick, J. A. (1995). Distribution of organochlorine pollutants in Atlantic sea turtles. *Copeia*, 2, 379-390.

Storelli, M. M., Barone, G. and Marcotrigiano, G. O. (2007). Polychlorinated biphenyls and other chlorinated organic contaminants in the tissues of Mediterranean loggerhead turtle Caretta caretta. *Science of the Total Environment,* 373, 456-463.

Storelli, M. M. and Marcotrigiano, G. O. (2000). Chlorobiphenyls, HCB, and organochlorine pesticides in some tissues of *Caretta caretta* (Linnaeus) specimens beached along the Adriatic Sea, Italy. *Bulletin of Environmental Contamination and Toxicology*, 64, 481-488.

Swarthout, R. F., Keller, J. M., Peden-Adams, M., Landry, A. M., Fair, P. A. and Kucklick, J. R. (2010). Organohalogen contaminants in blood of Kemp's ridley (*Lepidochelys kempii*) and green sea turtles (*Chelonia mydas*) from the Gulf of Mexico. *Chemosphere*, 78, 731-741.

Thompson, N. P., Rankin, P. W. and Johnston, D. W. (1974). Polychlorinated biphenyls and *p,p'*-DDE in green turtle eggs from Ascension Island, South Atlantic Ocean. *Bulletin of Environmental Contamination and Toxicology,* 11, 399-406.

Van de Merwe, J. P., Hodge, M, Olszowy, H. A., Whittier, J. M., Ibrahim, K.and Lee, S. Y. (2009). Chemical contamination of green turtle (*Chelonia mydas*) eggs in Peninsular Malaysia: implications for conservation and public health. *Environmental Health Perspectives*, 117, 1397-1401.

Van de Merwe, J. P., Hodge, M, Whittier, J. M., Ibrahim, K. and Lee, S. Y. (2010). Persistent organic pollutants in the green sea turtle *Chelonia mydas*: nesting population variation, maternal transfer, and effects on development. *Marine Ecology Progress Series*, 403, 269-278.

Zumbado, M., Goetahls, M., Álvarez-León, E. E., Luzardo, O. P., Cabrera, F., Serra-Majem, L. and Domínguez-Boada, L. (2005). Inadvertent exposure to organochlorine pesticides DDT and derivatives in people from the Canary Islands (Spain). *Science of the Total Environment*, 339, 49-62.

In: Pesticides: Characteristics, Uses and Health Implications ISBN: 978-1-62081-577-9
Editors: A. Abrego Gomez and E. Lugo de Ortega

Chapter 5

PESTICIDES IN THE ATMOSPHERE: MONITORING EMISSIONS AND ATMOSPHERIC DEGRADATIONS

L. A. Tortajada-Genaro[1] ***and E. Borrás***[2]
[1]Departamento Química, Universitat Politècnica de València, Spain
[2]Instituto Universitario Centro de Estudios Ambientales del Mediterráneo (CEAM-UMH), Spain

EMISSION OF PESTICIDES TO THE ATMOSPHERE

The agricultural production, in terms of quality, productivity,. and management, improved with the extensive application of pest control technologies, being today's agricultural practices heavily dependent on the use of them. The *pesticides* are the most important tool in intensive crop-based agriculture and in some public health applications. They are chemical or biological products used to prevent diseases, kill, repel, control plagues or interrupt the growth of plants, including herbicides, fungicides, acaricides, avicides, and insecticides. Different families of chemical compounds and formulations have been proposed to increase their efficacy using a wide range of action modes (Stoytcheva 2011). The most important families are organophosphates, carbamates, organochlorines, chloroacetanilines, anilines, chloroanilines, dinitroanilines, substituted ureas, and triazines, presented as solids, solutions, suspensions or emulsions. The pesticides can affect energy metabolism or growth in plants and nervous system, muscle action or growth regulation in animals. Unfortunately, the fraction of the dosage of applied pesticide enters into the different *environmental compartments* is quite important (*Bedos et al 2002*).

The distribution of a pesticide in the environment is controlled by their physical-chemical properties. For instance, Henry Law constant (the distribution coefficient between water and air) defines the predominant movement via air or via water phase (Bromilow 2005). The exposition risks come from pesticides present in the soil, waters, foods, and also the atmosphere. Therefore, their *adverse effects* on human health and the environment have changed the initial indiscriminate use of synthetic organic pesticides. Health effects in humans and other organisms are caused by skin contact, inhalation, and ingestion (Garrido-Frenich et al 2007).

The *atmosphere* represents the largest environmental compartment, one of the first destinations after the pesticide releases, and a relevant exposition source for living organisms. The human exposure to pesticides from the atmosphere is classified in occupational (farmers and related workers) and general population as the result of their high persistence or the exposition of degradation pesticide products. The accurate risk assessment is complicated because the adverse effects depend on factors as periods and levels of exposure (*Damalas et al 2011*). The characteristics of the atmosphere allow a fast movement, for instance the diffusion in air approximately 10^4 times faster than in water. There are several mechanisms for exchanging with the surface components (soils, water, plants or animals) through the boundary layer – lower area of troposphere in contact with terrestrial surface -.

The emission is the introduction of a pesticide into the atmosphere. When atmospheric contamination phenomenon occurs, it is a consequence of either a specific, identifiable source or place (point-source pollution) or it can be a combination of different sources in a wide area (non-point-source pollution). The emission of pesticides is produced during their manipulation, application or post-application (Table 1).

The reduction of the emission during the *manipulation* can be achieved with a correct use. The labeling of pesticide products should include warning statements that the pesticide handler should heed to avoid negative effects (Waxman et al 1998). Farmers or pest-control workers must take precautions to make sure that the pesticides remain in a sealed container until they are released into the application site, which also must be controlled to prevent the vapor from escaping or accidental spilling. Personal protective and application equipment requires a complete washing to reduce toxicological effects in workers, but these operations also produce the losses of pesticides to the indoor or outdoor atmosphere.

During the *application*, an important fraction of the pesticide always is lost and directly passes to the air. For instance, spraying by aircrafts, vehicle-mounted applicators, or manual equipments generate droplets with different sizes and some of them do not reach the target land surface or leaves. Losses during the application of pesticides can range to 50%, or even more in some cases (Van der Berg et al 1999). The main factors that affect pesticide air emission are the manner in which the pesticide is applied; concretely, the spray characteristics (e.g., size of nozzles), the equipment used (e.g., manual or vehicle), the formulation used (e.g., viscosity, volatility), the environmental conditions (e.g., wind speed, temperature, humidity, atmospheric stability), and finally the operators skills. Pesticides move away from the release site through wind (outdoor applications) or through air currents generated by ventilation systems (indoor applications).

The *post-application* emission has an important contribution. Even surface dissipation processes like wash-off due to rain or run-off are done, the atmospheric emission of pesticide are continuously produced until complete animal/plant uptake or soil adsorption (target or not-target). Their volatilization from the treated surface represents the most important input into the troposphere for days or even weeks after the application. Depending on the pesticide used and the climatic conditions, the volatilization can reach high percentage of the application dose. Vapor pressure, water solubility, and Henry's law constant could be considered the best parameters to describe the volatilization of a pesticide from applied surface or water (Unsworth et al 1999).

Table 1. Emission sources of pesticides into the atmosphere

Manipulation	Application	Post application
Losses during storage	Indoor application	Volatilization
Accidental spilling	Outdoor application	Wind erosion
Washing of application material		Chemical transformation
		Biological transformation
Waste materials		

So, volatilization from plants usually exceeds volatilization from soils due to adsorptive interactions. Meteorological conditions during the first days after the application are critical since most losses follow a diurnal pattern, with the largest losses at mid-day, when temperature and solar irradiation are higher. Field experiments carried out in different seasons confirmed that temperature highly affects the volatilization of the pesticides used (Ferrari et al 2003). Cultivation practices, moisture, and surface characteristics also affect the extent of volatilization. Another way in which pesticides can enter the atmosphere is via suspended particulate matter from wind erosion. The chemical transformations on target surfaces and the action of microorganisms also produce the breakdown of pesticide into components that can be emitted to the atmosphere.

ATMOSPHERIC PROCESSES

Pesticide movement away from the release site in the air is usually called drift. The intensive use increases significant contamination not only of the surroundings of the application areas but also of other remote areas. Nevertheless, these effects are result of several atmospheric processes which move out the pesticides of the use site and transform them (Figure 1). A measurement of its permanence is the *lifetime* of a pesticide in the atmosphere being the net result of physic-chemical removal processes.

The first process to consider after the emission to the atmosphere is the *dispersion* or the spontaneous tend of a pesticide to become well-mixed and diffused throughout the surface boundary layer. This dilution process favors the reduction of toxicological effects, but it also expands the area affected by a pesticide.

Simultaneously, the *gas-condensed phase distribution* is produced. Except some fumigants that are volatile compounds, pesticides are low-volatile or semi-volatile compounds, and they can therefore be distributed into gas and condensed phases. The coefficient K_i is the equilibrium constant that describes the gas-particle partitioning between free-gaseous form and particulate matter. This coefficient is the quotient of its concentration in particulate and gas phase. The distribution between the different phases is related to their physico-chemical properties, especially vapor pressure.

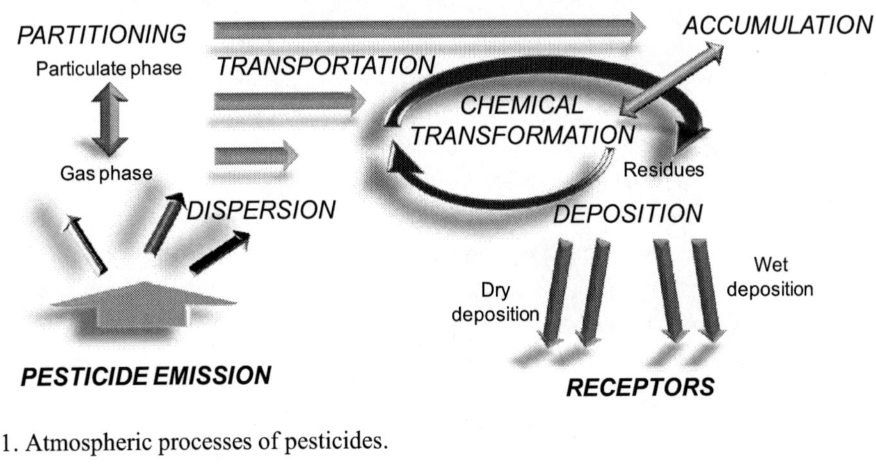

Figure 1. Atmospheric processes of pesticides.

Vapor pressures higher than 10^{-4} Pa imply that the compounds are mainly observed in the vapor phase, whereas vapor pressures lower than 10^{-6} Pa are almost always present in the particle phase (Huskes et al 1997). Nevertheless, once compound have begun to condense and a layer has formed on the existing seed particles, a portion of their mass is distributed into this condensed phase, even gas-phase concentrations are below their saturation concentrations (Odum et al. 1996). Other factors are the manner in which the pesticide is applied or the meteorological conditions during the application, because the distribution can shift.

Pesticides are moved due to meteorological processes from microscale (e.g., local winds) to macroscale (e.g., continental phenomena) (Bidleman 1999). The *transportation* of a pesticide depends on their nature (gaseous or particulate) and partly on the way they are released. An application close to the ground or floor is not as likely to be caught up in air currents as one from a greater height (upward direction or from an aircraft). On the other hand, the pesticide formulation also determines the transportation capability (Waxman 1998). Lightweight particles, such as dusts and wettable powders, or small spray droplets are easily carried by moving air. High-pressure and fine nozzles produce very small spray droplets that are very likely to drift. Meanwhile, granules, pellets and coarse nozzles are much heavier and tend to settle out of air quickly. Nevertheless, the chemical composition is the most important variable. For instance, low volatile pesticide is integrated in the particulate phase and it will have long-range transport capability only if the size of their aerosol particles is below 1 μm. There are several models used world-wide for describing the transport of pesticides through the air (*FOCUS 2008*). The models provide differences in the fate and spatial distribution and study temporal evolution of concentration levels from a dispersion scheme with 3-D wind, including variables such as turbulence, atmospheric stability or emission rates. But, the actual distance traveled by a pesticide strongly depends on the time it resides in the atmosphere. A pesticide with a short lifetime only affects the region near from the application site (local and regional scale). Meanwhile, pesticides with lifetimes longer than 2 days, called persistent pesticides, travel far from their emission point (large scale) and they can be lifted into the troposphere above the atmospheric boundary layer. So, the detection of pesticides has been described in remote areas without agricultural sites (e.g. mountain areas). Persistent pesticides can be transported even across countries, in fact, forbidden ones have been detected due to the application in neighbor countries where their application is not exhaustively controlled (*FOCUS 2008*).

A relevant removal mechanism of pesticides is the *chemical transformations* that can occur into the atmosphere. They are ozonolysis, photolysis or photochemical reactions with other pollutants in the atmosphere, as it is described in the following section. The result is the degradation of pesticide reducing its concentration present in the atmosphere, but new products are also formed. These secondary pollutants include gaseous or condensed compounds and they have a different toxicity, atmospheric residence time, and chemical stability than the original pesticide. The formation and growth of secondary particles are an example of chemical conversion. These chemical transformations are affected by the meteorological conditions as well as the physico-chemical properties of the pesticides. The degradation processes ranges from less than a minute to several years. The process in which persistent pesticides or their residues stay in the environment without breaking down for long periods of time is called *accumulation*. These molecules have desirable properties as long term pest control reducing the need for repeated applications, but there is also a higher likelihood contaminate remote places due to their low atmospheric chemical reactivity.

The main route of pesticide removal from air is deposition (Bidleman 1999). *Wet deposition* is the uptaking of the pesticide at the earth´s surface by precipitation, mainly in rain but also in fog and snow. Wet deposition is determined by the precipitation rate, the cloud, and rain drop distribution, the Henry's law constant, the air/water partition coefficient, and the washout ratio for particles. *Dry deposition* is the removal of particles and gases by sedimentation, diffusion or impaction processes, and their subsequent physical/chemical attachment to terrestrial surfaces (water, soil, and vegetation) or by respiration. This process is affected by the nature of the receiving body, being the most important parameters the Henry´s law constant, the partitioning coefficients of the pesticide in octanol/water, soil organic carbon/water, and octanol/air, the leaf area index, and the organic carbon content of the soil (Duyzer et al 1997).

So, larger-sized particles have only a short-range transport capability because sedimentation controls their removal. Pesticides adsorbed to small aerosols are found mainly in wet deposition, while pesticides in the vapor phase are probably more evenly divided between wet and dry deposition. Therefore, the major route for the deposition is by wet deposition via rainwater (Unsworth et al 1999). Although the deposition finishes, the pesticides could be re-emitted into the atmosphere under certain conditions and the cycle starts again.

In summary, after the emission the pesticides to the atmosphere, they are dispersed, transported, and transformed by various time-dependent physical and chemical processes (Figure 1). Pollutants spread in a three-dimensional flow, mixing with other atmospheric components, being these processes highly dependent on the atmospheric conditions. Simultaneously, the pesticides react to form secondary pollutants and they are deposited through several mechanisms, thus reducing the atmospheric levels.

CHEMICAL TRANSFORMATIONS

The behavior of a pesticide in the atmosphere is the same as other volatile and semivolatile organic compounds, that it is the degradation by reactions with other components of atmosphere (Atkinson et al 1999). A *residue* is the product of this breakdown process.

The number, structure, and concentration of the residues depend on atmospheric conditions. The residues are generally at a higher oxidation state, more polar, and better soluble in water, thus harmless components. But the degradation does not always result in a detoxification of the initial compound released. In the same way, the lifetime of residues in the atmosphere, or in the environment in general, can be totally different to the parent pesticide, since their new chemical properties can accelerate or reduce the removal processes. Nevertheless, it is said that a specific residue (or more) with a characteristic part of a pesticide or a pesticide family usually remains in the environment for a long period of time.

The *degradation rate*, the change in the concentration of the pesticide with the time, depends primarily on the chemical structure and it also may be affected by environmental conditions at the release site, such as surface type, chemical composition, and pH, surface moisture, temperature, exposure to direct sunlight (Waxman 1998). A slow degradation rate increases the probability of animals and humans can be harmed when they respire or they feed on plants that have deposited pesticide on them. The reaction rate is described for the following general expression of rate law, where k_R is the rate constant for the reaction of pesticide at concentration *[Pest]* and the reagent at concentration *[R]*.

$$rate = k_R [Pest] [R]$$

Most transformation occurs in the atmospheric boundary layer and troposphere, although persistent pesticide residues can reach the higher levels of the atmosphere, i.e. the stratosphere. Reactions can occur in the gas, the liquid or in the particle phase (Figure 2), and they follow several reaction pathways (Finlayson-Pitts et al 2000; Atkinson et al 1999). For instance, the initial partitioning between the gas and particle phases significantly influences the relative importance of the plausible transformations.

The *degradation routes* of pesticides in the atmosphere involve photolysis (also called direct photolysis) and/or reactions with ozone, OH, and NO_3 radicals (also called indirect photolysis). Reaction with gaseous nitric acid can be possible in urban areas where its concentration is significant. For nighttime, the main reactions are oxidations (i.e., ozone); meanwhile daytime reaction experiments are photolysis and photo-oxidations (i.e., OH). The main reactions are described in the following paragraphs.

The *photolysis* is a chemical process by which molecules are broken down into smaller units through the absorption of sunlight. The decomposition or isomerization reactions occur when the pesticide absorbs radiation at wavelengths between 290 nm and 800 nm. These limits correspond to the cut-off region of sunlight UV radiation and the latter wavelength is the longest one which can break a chemical bond, respectively. Photolysis reaction rate constants for pesticide degradations (*J*) range from 10^{-6} to 10^{-3} s^{-1}.

$$\text{Pesticide} + \text{sunlight} \rightarrow \text{Products}$$

The reaction with *hydroxyl radicals* (OH) is the main degradation reaction for a large number of pesticides. The OH radical is the most important reactive species in the troposphere; it reacts with all organic compounds, including pesticides (Atkinson et al 1999). This radical is formed during daylight hours mainly by the photolysis of O_3, nitrous acid, nitrogen oxides, hydrogen peroxide or formaldehyde. The OH concentrations are highest from the late morning to mid-afternoon, barely detectable at night, 4-20 times higher in summer than winter, and 2-3 time higher at 30^0N latitude compared to 60^0N (Bidleman 1999). Reaction rate constants for pesticide reaction with OH radicals (k_{OH}) range from 10^{-14} to 10^{-10} cm^3 $molecule^{-1}$ s^{-1}. The rate constants are 1.5-1.7 times higher at 25°C than 0°C.

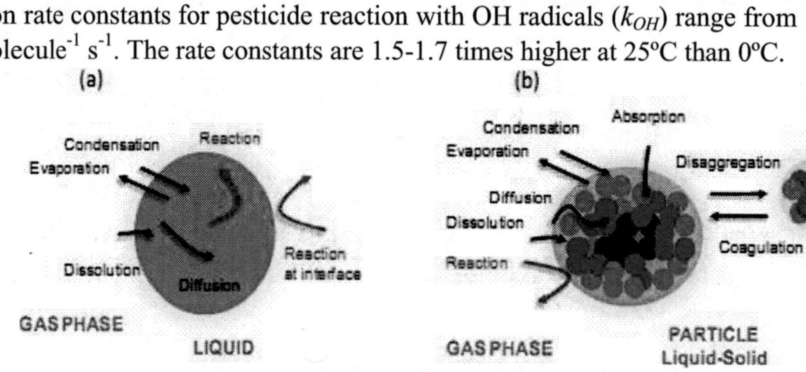

Figure 2. Physical process and reactions in liquid phase (a) and particles (b).

The main routes involving OH radicals are the abstraction of reactive hydrogen atoms, addition to non-aromatic multiple bonds, addition to aromatic rings and reactions with nitrogen, sulfur and phosphorus moieties (Unsworth et al 1999). The sequence of transformation processes for the abstraction of reactive hydrogen atoms starts with formation of an alkyl radical. This radical then reacts with oxygen to give an alkylperoxy radical which can yield the corresponding aldehyde or ketone.

Further reactions will give CO_2, nitrogen dioxide as well as mono and di-carboxylic compounds. The addition of OH radicals to aromatic rings results in the formation of hydroxyaromatic compounds (ring-retained compounds), hydroperoxy radicals or cleavage of the aromatic ring (ring-cleavage compounds).

The *ozonolysis,* or reaction with ozone (O_3), can be important for pesticides with activated unsaturated carbon-carbon bonds. Ozone arrives in the troposphere by diffusion from the stratosphere, and it is also formed in the troposphere from the reaction of organic molecules and nitrogen oxides in the presence of sunlight. Reaction rate constants for pesticide reaction with ozone (k_{O3}) range from 10^{-20} to 10^{-19} cm^3 $molecule^{-1}$ s^{-1}.

The *reaction with nitrate radical* (NO_3) is only important during night-time ad for pesticides containing double carbon-carbon bonds. NO_3 is formed in the troposphere by the reaction of NO_2 with O_3. Reaction rate constants for pesticide reaction with nitrate (k_{NO3}) range from 10^{-15} to 10^{-14} cm^3 $molecule^{-1}$ s^{-1}.

The parameter to describe the reactivity of a pollutant in the atmosphere is *atmospheric lifetime* (τ), calculated as the ratio between the time needed for a pesticide to disappear 1/e of its initial concentration.

A high τ value means that the pesticide or residue persists in the atmosphere without degradation. The lifetime for each pesticide is related to the reaction rates of all degration processes that participates and it is calculated from the expression:

$$\tau = \frac{1}{J} + \frac{1}{k_{OH}[OH]} + \frac{1}{k_{O3}[O_3]} + \frac{1}{k_{NO3}[NO_3]}$$

where J, k_{OH}, k_{O3}, and k_{NO3} are the rate constants for the photolysis, reaction with hydroxyl radicals, ozone, and nitrate radicals respectively, and $[OH]$, $[O_3]$, and $[NO_3]$ are the correspondent molar concentrations.

In general terms, organic pesticides have a lifetime in air between few minutes for reactive pesticides and several days for persistent pesticides. For instance, considering [OH] = 2×10^6 molecule cm^3 (average 12-hour daytime concentration), lifetime of trifluralin is 15 minutes (LePerson et al 2007), 2 hours for chlorpyrifos (Hebert et al 2000), 3.5 hours for chlorpyrifos-methyl (Muñoz et al 2011), 5.4 hours for chloropicrin (Vera et al 2010), and 304 days for lindane (Brubaker and Hites 1998). Residues are generally more stable products with lower degradation rates and they have considerably longer atmospheric lifetimes than parent pesticides.

Furthermore, as pesticides travel from temperate to polar regions, their atmospheric lifetimes will increase due to lower rate constants and fewer radicals. This fact explains the detection in arctic air of pesticides which have estimated lifetimes of a few days under temperate conditions (Bidleman 1999).

MONITORING OF ATMOSPHERIC DEGRADATIONS

Several methods have been used to determine the *volatilization of pesticides* after their application into the soil, such as the aerodynamic-gradient method, a theoretical profile method, the integrated horizontal flux method, and the box method (Van den Berg et al 1999). For pesticides which are sprayed in crops, the rate of volatilization has been measured with methods, such as aerodynamic-gradient, the integrated horizontal flux method or sine accumulation methods. Field volatility experiments are costly and volatilization chambers and wind tunnel systems have been developed to simulate applications. In these systems, weather variables such as wind speed, air temperature, humidity, and solar radiation can be controlled.

The *degradation of pesticides in liquid phase* has been studied by several research groups (Barcelo et al 1997; Ikehata et al 2006; Martínez Vidal et al 2007). These experiments tried to reproduce the reactions controlled by light in methanol, hexane or aqueous solutions, or on soil or leaf surfaces.

Despite its importance, there are only few studies on *degradation in air*. Moreover, experimental data has been reported for a limited number of pesticides (*FOCUS 2006*). This lack of information is a consequence of the complexity of the chemical processes involved in pesticide emissions and the difficulties associated with having to carry out the laboratory studies under realistic conditions. These problems increase when the objective is the analysis or the characterization of degradation products. In fact, few reaction products observed from gaseous and particulate phases are available (Atkinson et al 1999; Vera et al 2010; Muñoz et al 2011). This is the case of DDEs, as products of the degradation of DDTs; phosgene, that can come from the photolysis of chloropicrin or from the photoxidation of dichlorvos, oxones from the organothiophosphate or organophosphorous pesticides, and p-nitrophenols from dinitroanilines. The rate constants and mechanisms of several reactions with ambient oxidants are still uncertain. Studies are carried out based on the following research strategies (Unsworth et al 1999), including experimental estimation methods, laboratory tests to measure values for the respective reaction constants, and experiments in outdoor chambers and/or field studies.

The most often used technique for obtaining data on the transformation rates of pesticides is the use of *Structure Activity Relationships* (SAR). The method involves a standard group rate constant for a particular reaction, which is then adjusted depending on the relevant substituents (Kwok and Atkinson 1995). The result is the transformation rate for the target compound based on dominant reaction paths. The calculation is made by programs such as the Atmospheric Oxidation Program. So, the approach for reactions with OH radical further assumes that there is a number of different reaction pathways and that the overall rate constant is equal to the sum of these interactions. The degradation rate constant can then be determined using a mean OH radical concentration in air, for example a 0.5×10^6 OH radicals cm^{-3} for a 24 hour day or 1.5×10^6 OH radicals cm^{-3} for a 12 hour day. Estimations of rate constants were compared to those measured at room temperature in the gaseous phase (*FOCUS 2006*). Acceptable correlations were obtained for compounds with relatively simple low molecular weight and a limited number of functional groups. For larger, more complex pesticides, the program seems overestimate reaction rates.

A better experimental approach to studying the atmospheric transformations involves the use of *atmospheric reactors*. These consist of reaction chambers with a light source to generate radicals, analytic devices, and a gas line for introducing the pesticide and reactants into the bags. However, for representative results, a large reaction vessel is required. Large chambers minimize chamber surface area-to-volume ratios, thereby reducing wall effects such as the adsorption and desorption of gas-phase species or particle compounds. In this sense, the photochemical reactors or 'smog chambers' are invaluable tools in the investigation of atmospheric sciences (Cocker et al 2001, Barnes et al 2006, Nilsson et al 2009). The simulators allow the examination of specific atmospheric chemical transformations, since precise oxidant conditions can be reproduced. The photochemical reactor applications include determinations of lifetimes and relative rates of reaction, studies on product distributions, and development of degradation mechanisms. An example of a full-equipped photo-simulation chamber is the European Photoreactor (EUPHORE). This facility, with two photoreactors (200 m^3), is currently one of the largest and best-equipped outdoor simulation chambers in the world (Borrás et al. 2011a).

The basic elements of a *photoreactor* are the chamber structure, the control system, the light source (natural or artificial), the injection/formation modules of reactants, the pesticide injection system, the mixing components, and the monitoring instruments (Becker 1996; Cocker et al 2001). Several measurements systems are integrated for registering physical parameters (radiation, pressure, humidity, and temperature) and monitoring chemical parameters such as precursor species, intermediate, and final products. For all of that, several instruments are integrated. The JNO_2 value of light is measured with a spectral radiometer. Pressure, humidity, and temperature are measured by means of specific sensors. Most photoreactors are equipped with a white-type mirror system coupled to a Fourier Transform Infrared spectrometer with a long absorption path length, and a high spectral, and time resolution for determining gaseous products. Simultaneously, different monitors of the main atmospheric gases are integrated for measuring in the low concentration ranges (NO_x, O_3, SO_2, CO, CH_4, HCHO, HONO, etc). The laser-induced fluorescence instrument allows the monitoring of OH and HO_2 radicals. Full-equipped smog chambers have several on-line analytical instruments for measuring gaseous compounds: gas and liquid chromatographs with several detectors, differential optical absorption spectrometer, and proton transfer resolution – mass spectrometer. For aerosol size characterization two types of instruments are used: those that provide size distribution data and those that collect samples for subsequent analyze such as gravimetric, chemical or microscopic analysis. These instruments provide information on physical properties such as size distributions, aerosol yielding, and concentration profiles. In addition, the integration of cartridges or filter systems allows a selective sampling to be made of specific families of gaseous or condensed compounds for subsequent analysis.

A typical *experiment in a photoreactor* consists of studying a specifically induced reaction of pesticide with atmospherically relevant radicals under controlled conditions, isolated from meteorological variations or dispersion. For this, the photoreactor is filled with air in the adequate composition - fixed concentrations of reactants- and mixed with pesticide under a time-controlled light exposition. The steps in a large photoreactor experiment are: set-up of chamber facility, background determinations, reaction monitoring, and chamber cleaning. For nighttime experiments (oxidations), the start of the reaction is considered to occur when all components are mixed. For daytime reaction experiments (photolysis and photo-oxidations), the chamber is generally opened (outdoor chambers) or the lights are switched on (indoor chambers) after mixing.

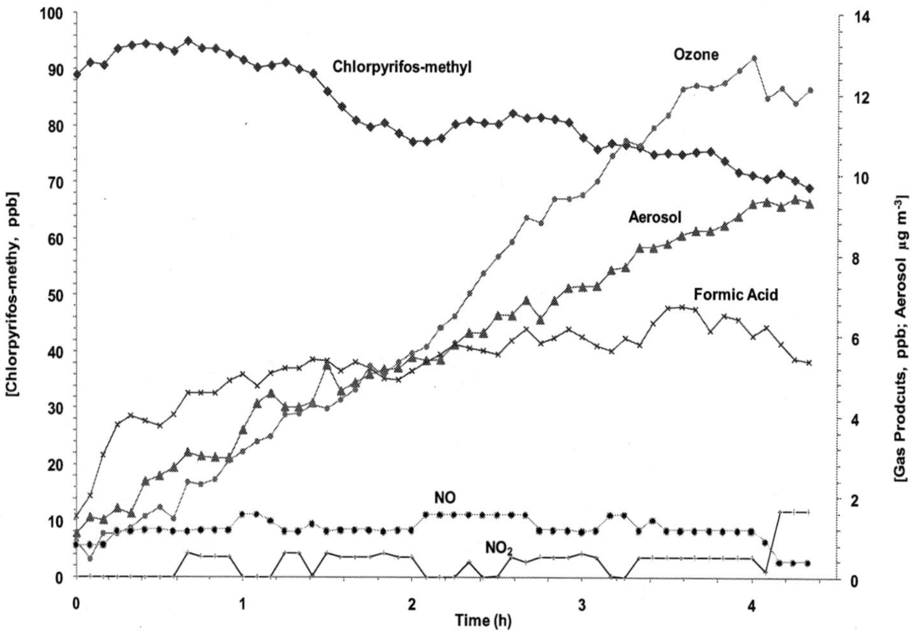

Figure 3. Degradation profile of chlorpyrifos-methyl. Reactions conditions: photolysis in outdoor chamber of 200 m^3.

The experiment duration ranges between 0.5-48 h, according to the following criteria: the aerosol level or the formation of a selected product is reached, all the reactants are consumed, the light exposition time is exceeded or aging of relevant species is studied. Figure 3 shows an example of experimental data collected in a degradation experiment performed in an outdoor atmospheric chamber.

The lifetimes of pesticide emissions are determined from the corresponding measured reaction rate constants of each individual process. Some examples of the degradation of pesticides in air are the studies on the degradation of tertbutylazine (Palm et al 1997), chlorotoluron and isoproturon (Millet and Zetzsch 1998), dichlorvos (Feigenbrugel et al,2006), trifluraline (Le Person et al 2007), chloropicrin (Vera et al 2010), chlorpyrifos-methyl (Muñoz el al 2011), and hymexazol (Vera et al 2011).

Even so, the implications for human health and the environment of pesticides should produce more studies about the atmospheric degradation in air. The research efforts may be especially focused on degradation products, since they have been considerably longer atmospheric lifetimes and consequently long-range transport.

Techniques for Pesticide Analysis in Atmospheric Samples

The concern about environmental and health risks has raised the regulations about the approbation of pesticides and the standards of safety required. Surveillance programs complete the strategies to guarantee the protection against their negative effects. Then, the information available on the monitoring of pesticides in the atmosphere is continuously increasing. The target samples are air –in the gas and/or particle phase-, rainwater, fog, and snow. During the past decades, the largest amount of data had been collected for pesticides in rainwater. Although the sampling strategy (raining collectors) is different, the analytical methods can be considered quite similar to those used in other environmental samples such as surface waters. Interesting review articles described the state-of-art about analytical techniques in rainwater samples and the pesticide concentrations in these samples (Van Dijk et al 1999; Dubus et al 2000; FOCUS 2006; Vogel et al 2008). Nowadays, a relevant analytical challenge is also detecting multi-residue pesticides at trace levels in gas and/or particle phase in air.

In air samples, the reported values are generally the sum of pesticides present in gas and particulate phases, even though some studies reported the distribution between gas and particles. Near 50% of the pesticides detected are herbicides being more than half nowadays forbidden according to the EU regulations (Yusa et al 2009). Concentrations of pesticides in ambient air ranged from few pg m^{-3} to many ng m^{-3}. The concentration levels show a seasonal behavior related to application periods. The highest concentrations in air usually occur in the spring and summer months coinciding with application times.

The highest pesticide concentrations are found in samples from the immediate surroundings of the application area (few kilometers scale). The exceptions are persistent pesticides or certain resides that they have been collected far away from the application site (Unsworth et al 1999; Coscolla et al 2010) and at other moments of the year after the application (Scheyer et al 2007). Most studies point to higher pesticide concentrations on agricultural/rural sites than in urban areas, as expected. Their presence is strongly related to transport from rural areas of significant agricultural importance, rather their application in urban pest control. However, a great number of pesticide residues can come from transformations induced for pollutants which are typically present in higher concentrations on urban sites.

Pesticides can be broadly classified according to their general chemical nature into several principal types as inorganic and organic (Nollet et al 2010). The analysis of *inorganic pesticides* is performed by well-established standard methods of trace element analysis. Meanwhile, the analysis of *organic pesticides* needs to be carried out on a variety of strategies to allow the determination of all families. At the same time, their analysis in air samples is often complicated due to sampling, extraction, cleanup or concentration problems. The analytical enforces are focused in the accurate, selective, and sensible determination of particulate phase and gaseous pesticides in ambient air. Recent reviews on pesticides in ambient air highlight the key requirements for new analytical methods (Yusa et al 2009; Kosikowska et al 2010).

The *sampling* in air samples is particularly interesting. Although a representative sample is quite easy to obtain, an appropriate sampling technique is necessary to achieve the sensitivity of the analytical instruments. The most common sampling techniques are grouped into two categories: passive and active samplers. *Passive samplers* are based on diffusion phenomena onto a material with a high-retention capacity. These systems require long sampling periods, ranging from few weeks to several months, and they are able to collect only gaseous phase pesticides. The retention medium utilized is polyurethane foam disks, semi-permeable membrane devices, polymer-coated glass samplers, and XAD resins. In *active samplers*, the air is pumped through a filter and/or a specific solid adsorbent. The pesticides in the particulate phase are retained in the filter, whereas those present in the gas phase are trapped by the adsorbent. High-volume samplers normally work a flow rate higher than 13 m^3

h^{-1} with sampling times ranging from 12 h to 7 days. Low-volume samplers (flow rates < 3.6 m^3 h^{-1}) are used for samples with high concentration levels, e.g., greenhouse air or occupational studies. Several filters and gaseous sampling media, from polyurethane foams to solid-phase cartridges, have been described previously. Recently, we have published a study comparing the efficient sampling of atmospheric pesticides depends on the chemical nature of the sorbent (Borrás et al 2011b). Thus, two types of filters – glass fiber and quartz fiber- for solid phase pesticides and four solid-phase sorbents – XAD-2, XAD-4, Florisil, and Orbo 49P – for gaseous phase pesticides were evaluated. The target analytes (16 pesticides) were selected from the most important authorized, banned or relatively recent banned classes of pest control agents based on spraying application. They included organochlorines, chloroacetanilines, anilines, chloroanilines, dinitroanilines, organophosphates and triazoles. Recovery percentages obtained are summarized in Figure 4.

In case of totally unknown samples, a first option is the *preliminary characterization* of pesticide residues by a screening method that provides qualitative information about the composition. The proposed methods are mainly based on thin-layer chromatography, enzymatic tests, and immunoassays in lateral low strip format. However, this strategy is difficult to apply in case of air samples.

The *extraction* of the residue from the sample matrix using an efficient and selective solvent is one of the most important steps in the analysis. The techniques are fluid-phase partitioning methods including Soxhlet extraction, liquid–solid extraction, pressurized fluid liquid extraction, and recently microwave-assisted extraction. Some sorptive based extraction methods (i.e., solid phase extraction, solid-phase microextraction, stir bar sorptive extraction) have been proposed.

The selection of pure and mixed organic solvents to be employed is governed by the type of pesticide under examination (polarity). A relevant method is liquid–solid extraction for pesticides trapped in filters and cartridges using an appropriate organic solvent. The solvent is removed to enrich the dissolved residues by using techniques such as freeze concentration, lyophilization, evaporation, distillation, reverse osmosis, and ultrafiltration.

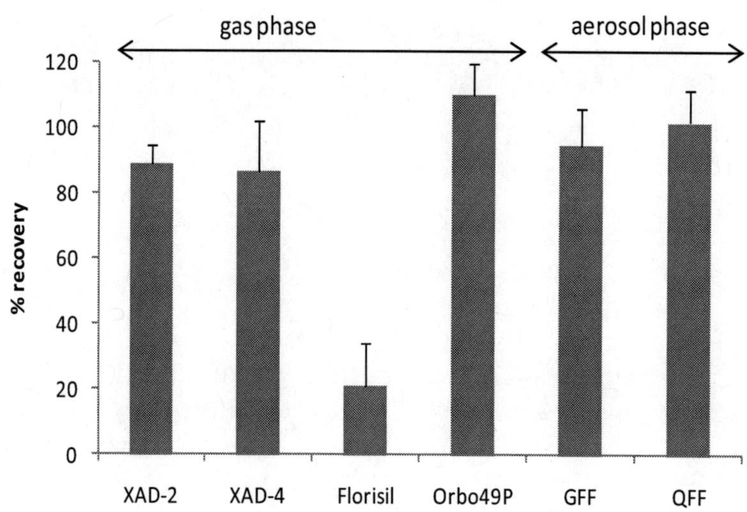

Figure 4. Retention efficiencies (%) of pesticides as function of adsorbent nature.

There are several pretreatment procedures for improving the extract before the determination. The *cleanup* is the removal of interfering substances from the extract, because air samples can contain a significant amount of other components that can be co-extracted together with pesticides.

The common approaches purify extracts by SPE columns with a specific composition covering a wide polarity range. The *enrichment* of the analyte is required in some cases due to the low concentration levels of residues in atmospheric samples. Concentration of pesticide residue can be below the lower limit of estimation of the available analytical technique; therefore, a suitable preconcentration/enrichment method is coupled with the analytical technique. The choice of enrichment method depends on factors such as the volatility and solubility of the pesticide residues, the degree of concentration required, and the nature of the analytical technique to be used.

The *determination* of the quantity of pesticide residues, together with metabolites and breakdown products has been described by several methods. An important group is bioanalytical methods, including bioassays, enzymatic measurement and bioreagent-based methods.

Bioassays measure the growth, death, or some other physiological change in animals, plants, or microorganisms. The main disadvantages are its lack of specificity and high detection limits. Enzyme-linked immunosorbent assays (ELISA) provides excellent results with minimal pretreatment of sample (Brun et al 2005). Biosensors are a low-cost tool to determine in environmental samples (Morais et al 2009). Spectrophotometric determinations are also relevant methods because they are simple, and inexpensive.

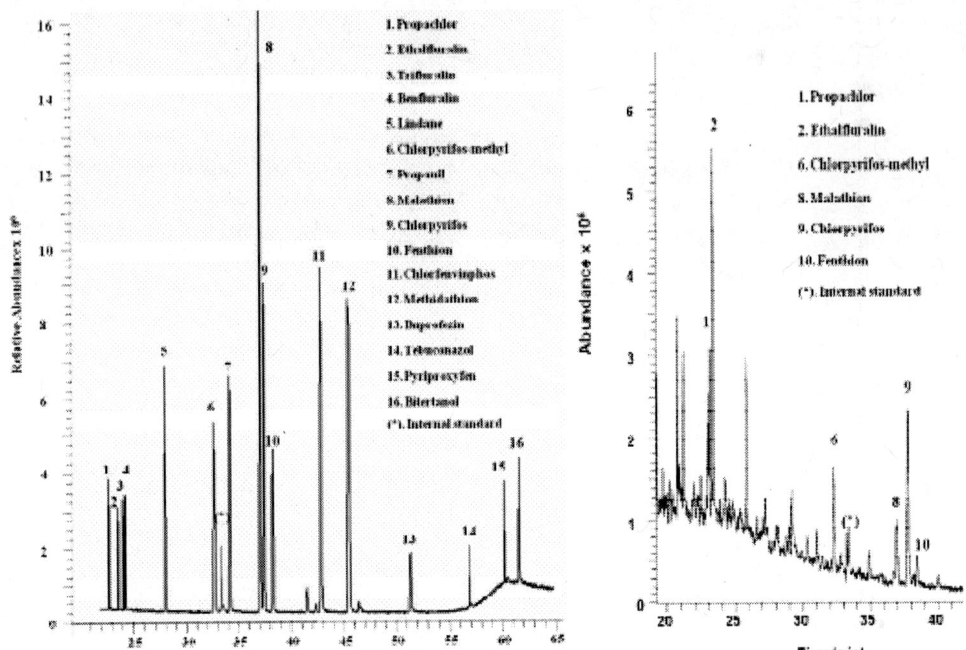

Figure 5. TIC chromatogram by splitless injection (a) Standard pesticide mixture at 1 mg L^{-1}; (b) Gaseous phase sample collected in an agricultural area (XAD-4 resin cartridge, sampling flow of 1 L min^{-1} for 24 h at 1-m above ground level).

Many of the chlorinated pesticides can be determined by colorimetric methods. Techniques as nuclear magnetic resonance or Infrared spectra have been used for the confirmation of pesticide residues. Nevertheless, the most important methods are based on chromatographic separation being extensively reported for the determination of pesticides in air samples. Gas-chromatography is generally used, except for some polar and sometimes thermal-labile compounds - e.g., some herbicides- being recommended liquid chromatography. Over the past decades, approaches to detect trace levels of pesticides have changed significantly, moving to chromatographic methods coupled to mass–spectrometry detection (Tadeo 2008; Yusa et al 2009; Borrás et al. 2011b). Its capacity to provide quantitative and confirmatory results is highly valuable for analyzing multiresidual pesticides. Figure 5 shows examples of GC-MS chromatograms obtained using the method described in reference Borrás et al. 2011b. Sixteen relevant pesticides from different chemical families were determined in less than one hour.

CONCLUSION

After six decades of the initial use of organic pesticides, the concern about environmental and health risks has raised the number of research studies focused on the determination of pesticides in atmosphere. They describe the physical and chemical processes after their emission and the strategies to guarantee the correct monitoring in atmospheric samples. The advances in scientific knowledge have shown pesticides mainly enter into the atmosphere during application and volatilization from plants and soil. Atmospheric process in the boundary layer will determine whether pollutants will be dispersed and diluted, transformed, exported to the free atmosphere or settled on the ground. So, a pesticide should be transported short or long distances depending on its residence time and the dominant weather conditions. Atmospheric degradations follow several reaction pathways and they occur in gas, liquid or particle phase. Nevertheless, direct photolysis and reaction with OH radicals are considered as the dominant transformations. Another important research area is the description of these reactions in air because their complexity limits the performance of relevant experiments. It has been shown that the better tool is the use of high-volume photochemical reactors. Analytical methodologies incorporated are capable of high resolution measurements at very low levels. Then, most of kinetic parameters and degradation products of pesticides have been obtained in this kind of facilities.

The last research area described in this chapter is related to the monitoring of pesticide residues in atmospheric samples. Different analytical techniques have been suggested for the detection, identification, and quantification. During the last decades the objectives of the investigations have increased. Studies related with pesticides in ambient air were mainly focused in emission of pesticides to the air during application in open or confined areas with high pesticide concentrations in order to define the working expositions. Nowadays, the monitoring of ambient air in remote areas at trace levels has gained in relevance.

Consequently, powerful detection techniques, such as MS coupled to GC or HPLC, are necessary tools in the modern analysis of pesticides in air. Sampling and sample preparation is still the analytical challenge, since otherwise interferences or matrix effect can occur. Despite of the advances, additional studies are required to reach a better knowledge of the occurrence, fate, and impacts in the atmosphere of the currently used pesticides or their residues.

ACKNOWLEDGMENT

The authors also thank J.T.B. for his contribution.

REFERENCES

Atkinson, R.; Guicherit, R.; Hites, R.A.; Palm, W.U.; Seiber, J.N.; de Voogt, P. Transformations of pesticides in the atmosphere: A state of the art. *Water, Air, Soil Pollut.* 1999, 115, 219-243.

Barceló, D.; Hennion, H.C. Trace determination of pesticides and their degradation products in water. Elsevier. 1997.

Barnes, I.; Rudzinski, K.J. Environmental simulation chambers: application to atmospheric chemical processes. *NATO Science Series Dordrecht*, The Netherlands. Springer. 2006.

Becker, K.H. Design and Technical Development of the European Photoreactor and First Experimental Results, *Final Report of the EC-Project (EV5V-CT92-0059),* Wuppertal. 1996.

Bedos, C.; Cellier, P.; Calver, R.; Barriuso, E.; Gabrielle, B. Mass transfer of pesticides into the atmosphere by volatilization from soils and plants: overview. *Agronomie.* 2002, 22, 21-33.

Bidleman, T. F. Atmospheric transport and air-surface exchange of pesticides. *Water, Soil and Air Pollut.* 1999, 115, 115-166.

Borrás, E.; Tortajada-Genaro, L.A. Simulator photoreactor for the chemical characterization of diesel exhaust emissions (Chapter 12) in *Advances in Energy Research.* Volume 2. ed. Nova Science Publishers. Hauppauge NY. 2011a

Borrás, E.; Sánchez, P.; Muñoz, A.; Tortajada-Genaro, L.A. Development of a gas chromatography–mass spectrometry method for the determination of pesticides in gaseous and particulate phases in the atmosphere. *Anal. Chim. Acta* 2011b, 699, 57– 65.

Bromilow, R.H. *Pesticides. Encyclopedia of Soils in the Environment*, 2005, Pages 188-195.

Brubaker, W.W.; Hites, R.A. OH reaction kinetic of gas-phase α- and γ-hexachlorocyclohexane hexachlorobenzene. *Environ. Sci. Technol.* 1998, 32, 766-769.

Brun, E.M.; Garces-Garcia, M.; Banuls, M.J.; Gabaldon, J.A.; Puchades, R.; Maquieira, A. Evaluation of a novel malathion immunoassay for groundwater and surface water analysis. *Environ. Sci. Technol.* 2005, 39, 2786-2794.

Cocker, D.R.; Flagan, R.C.; Seinfeld, J.H. State-of-the-art chamber facility for studying atmospheric aerosol chemistry. *Environ. Sci. Technol.* 2001, 35, 2594–2601.

Coscolla, C.; Colin, P.; Yahyaoui, A.; Petrique, O. Yusa, V.; Mellouki, A.; Pastor, A. Occurrence of currently used pesticides in ambient air of Centre Region (France) *Atmos. Environ.* 2010, 44, 3915-3925.

Damalas, C.A; Eleftherohorinos, I.G. Pesticide exposure, safety issues, and risk assessment indicators. *Int. J. Environ. Res. Public Health* 2011, 8, 1402-1419.

Dubus, I.G.; Hollis, J.M.; Brown, C.D. Pesticides in Rainfall in Europe, *Environ. Poll.* 2000, 110, 331-344.

Duyzer, J.H.; Van Oss, R.F. Determination of deposition parameters of a number of organic pollutants by laboratory experiments. *TNO report* R97/150, Apeldorn, The Netherlands. 1997.

Feigenbrugel, V.; Le Person, A.; Le Calve, S.; Mellouki, A.; Munoz, A.; Wirtz, K. Atmospheric fate of dichlorvos: Photolysis and OH-initiated oxidation Studies. *Environ. Sci. Technol.* 2006, 40, 850-857.

Ferrari, F.; Trevisan, M.; Capri, E. Predicting and measuring environmental concentration of pesticides in air after soil application. *J. Environ. Qual.* 2003, 32, 1623-1633.

Finlayson-Pitts, B.J.; Pitts, J.N. *Chemistry of the Upper and Lower Atmosphere: Theory, Experiments and Applications*. Academic Press, San Diego, CA. 2000.

FOCUS Working Group, *Pesticides in Air: Considerations for Exposure Assessment*, European Union, Brussels, SANCO/10553/2006 Rev, June 2, 2008.

Garrido-Frenich, A.; Egea-González, F.J.; Marín-Juan, A.; Martínez-Vidal, J.L. Hazard identification and human exposure to pesticides In: *Thermodynamics, Solubility and Environmental Issues* T.M. Letcher (ed). Elsevier. 2007.

Hebert, V.R.; Hoonhout, C.; Miller, G.C. Reactivity of certain organophosphorus insecticides toward hydroxyl radicals at elevated air temperatures. *J. Agric. Food Chem.* 2000, 48, 1922-1928.

Huskes, R.; Levsen, K. Pesticides in rain, *Chemosphere,* 1997, 35, 3013-3024.

Ikehata K.; El-Din M.G. Aqueous pesticide degradation by hydrogen peroxide/ultraviolet irradiation and Fenton-type advanced oxidation processes: a review. *J. Environ. Eng. Sci.* 2006, 5, 81-135.

Kwok, E.S.C.; Atkinson, R. Estimation of hydroxyl radical reaction-rate constants for gas-phase organic-compounds using a structure-reactivity relationship - an update. *Atmos. Environ.* 1995, 29, 1685-1695

Kosikowska, M.; Biziuk, M. Review of the determination of pesticide residues in ambient air. *Trends Anal. Chem.* 29 (2010) 1064–1072.

LePerson, A.; Mellouki, A.; Muñoz, A.; Borras, E.; Martin-Reviejo, M.; Wirtz, K. Trifluralin: Photolysis under sunlight conditions and reaction with HO radicals. *Chemosphere.* 2007, 67, 376–383.

Martínez Vidal, J.L.; Plaza-Bolaños, P.; Romero-González, R.; Garrido Frenich, A. Determination of pesticide transformation products: A review of extraction and detection methods. *J. Chromatogr. A* 2009, 1216, 6767-6788.

Morais, S.; Tortajada-Genaro, L.A.; Arnandis-Chover, T.; Puchades, R.; Maquieira, A. Multiplexed Microimmunoassays on a Digital Versatile *Disk. Anal. Chem.* 81 (2009) 5646–5654.

Muñoz, A.; Vera, T.; Sidebottom, H.; Mellouki, A., Borrás, E.; Ródenas, M. Clemente, E.; Vázquez, M. Studies on the Atmospheric Degradation of Chlorpyrifos-Methyl. *Environ. Sci. Technol.* 2011, 45, 1880–1886.

Nilsson, E. J.K.; Eskebjerg, C.; Johnson, M.S. A photochemical reactor for studies of atmospheric chemistry. Atmos. Environ. 2009, 43, 3029-3033.

Nollet, L.M.L.; Rathore, H.S. (ed.). *Handbook of pesticides*. Methods of pesticide residues analysis CRC Press, 2010.

Odum, J.R., Hoffmann, T., Bowman, F.M., Collins, D., Flagan, R.C., Seinfeld, J.H. Gas/particle partitioning and secondary organic aerosol yields. *Environ. Sci. Technol.* 1996, 3, 2580–2585.

Scheyer, A.; Morville, S.; Mirabel, P.; Millet, M. Variability of atmospheric pesticide concentrations between urban and rural areas during intensive pesticide application. *Atmos. Environ.* 2007, 41, 3604-3618.

Stoytcheva, M. (ed.) *Pesticides - Formulations, Effects, Fate*. InTech Publisher, 2011.

Tadeo J.L. (Ed.), *Analysis of Pesticides in Food and Environmental Samples*, CRC Press, Boca Raton, FL, 2008.

Unsworth, J. B.; R. D. Wauchope; A. W. Klein ; E. Dorn; B. Zeeh; S. M. Yeh; M. Akerblim; K. D. Racke, and B. Rubin, 1999. Significance of long range transport of pesticides in the atmosphere. *Pure Appl. Chem.*, 71, 1359-1383.

Van Den Berg, F.; Kubiak, R.; Benjey, W.G.; Majewski, M.S.; Yates, S.R.; Reeves, G.L.; Smelt, J.H.; Van Der Linden, A.M.A. Emission of pesticides into the air. *Water, Air and Soil Pollut.* 1999, 115, 195-218.

Van Dijk, H.F.G. and Guicherit, R. Atmospheric Dispersion of Current-Use Pesticides – A Review of the Evidence from monitoring studies. *Water, Air, Soil Pollut.* 1999, 115, 21-70.

Vera, T.; A. Muñoz; M. Ródenas; M. Vázquez; A. Mellouki; J. Treacy; I. Al Mulla; H. Sidebottom. Photolysis of trichloronitromethane (chloropicrin) under atmospheric conditions. *Z. Phys. Chem.* 2010, 224, 1039-1057.

Vera, T.; Munoz, A.; Rodenas, M.; Vazquez, M.; Borras, E.; Marques, M.; Mellouki, A.; Treacy, J.; Sidebottom, H. Atmospheric fate of hymexazol (5-methylisoxazol-3-ol): Simulation chamber studies. *Atmos. Environ.* 2011, 45, 3704-3710.

Vogel, J.R.; Majewski, M.S.; Capel, P.D. Pesticides in Rain in Four Agricultural Watersheds in the United States. *J. Environ. Qual.* 2008, 37, 1101-1115.

Waxman, M.F. *Agrochemical and pesticide safety handbook*. Lewis publishers, 1998.

Yusà, V.; Coscollà, C.; Mellouki, W.; Pastor, A.; de la Guardia, M. Sampling and analysis of pesticides in ambient air. *J. Chromatogr. A.* 2009, 1216, 2972-2983.

In: Pesticides: Characteristics, Uses and Health Implications ISBN: 978-1-62081-577-9
Editors: A. Abrego Gomez and E. Lugo de Ortega

Chapter 6

PESTICIDES: INDIVIDUAL TO ECOSYSTEM LEVEL EFFECTS

Azad Mohammed
Department of Life Sciences,
The University of the West Indies, Trinidad and Tobago

WHAT ARE PESTICIDES?

Pesticides are among the most widely used chemicals in the world, and are also one of the most dangerous to human health. According to the Stockholm Convention on Persistent Organic Pollutants, 13 of the 21 most dangerous and persistent organic chemicals are pesticides. The Food and Agriculture Organization (FAO) has defined pesticides as: "*any substance or mixture of substances intended for preventing, destroying or controlling any pest, including vectors of human or animal disease, unwanted species of plants or animals causing harm during or otherwise interfering with the production, processing, storage, transport or marketing of food, agricultural commodities, wood and wood products or animal feedstuffs, or substances which may be administered to animals for the control of insects, arachnids or other pests in or on their bodies. The term includes substances intended for use as a plant growth regulator, defoliant, desiccant or agent for thinning fruit or preventing the premature fall of fruit. Also used as substances applied to crops either before or after harvest to protect the commodity from deterioration during storage and transport.*"

Pesticides are designed to kill, suppress or alter the life cycle of pests, such as insects, rodents, fungi and unwanted plants (weeds). In the public health sector, they are used to control vectors of diseases while in agriculture they are used to kill pests. However, they are not selective, and can also kill or cause harm to non-target organisms. The most common types of pesticides include: herbicides, insecticides, fungicides, rodenticides, pediculicides and biocides.

Pesticides generally contain many different molecules including:

- An active ingredient (AI) which is responsible for the potency of the chemical (about 900 active ingredients are currently available, e.g. acetamiprid, diazinon, endosulfan)

- A solvent carrier for the active ingredient (e.g. Isophorone)
- Additives which have little or no biological activity, but can modify the qualities of the pesticide (e.g. surfactants, antifoams).

Presently, there are over 100,000 pesticides available on the global market. Global usage has increased steadily over the past 60 years, averaging about 2.43 million metric tons (Active ingredients) annually. The highest usage occurs in the Americas and Europe (Table 1), with America, France and Japan being the largest consumers. However, many countries often report reduction in usage, but it is essential to remember that the current active materials are much more effective than those of the previously available.

The increased pesticide usage has occurred because of the greater demands for agricultural products as countries seek to ensure food security and meet the needs of increasing populations. However, this has resulted in significant impacts in terrestrial, aquatic and atmospheric environments.

Table 1. Estimated global consumption of pesticides (FAO 2009)

Geographical region	Estimated total consumption (metric tons)	Year
Africa		
Eastern Africa	2228	2009
Middle Africa	6601.4	2009
Northern Africa	1040.8	2009
Western Africa	15306.5	2009
Americas		
United States	398636.4	2001
Canada	29206	1994
South America	58537.5	2009
Asia		
Central Asia	245.2	2009
Eastern Asia	72400.6	2009
Southern Asia	4884.7	2009
South Asia	76469.4	2009
Western Asia	2442.1	2009
Europe		
Eastern	37296.1	2009
Northern	34857.6	2009
Southern	39795.1	2009
Western	72731.9	2009

Pesticides pose varying levels of risk in the environment as exposure can occur via different routes; dermal, inhalation, oral, ingestion and through ocular contact. Pesticides in the environment can be insidious for three reasons. First, they can become concentrated through successive trophic levels. The highest concentrations often occur in the biomass of top predators.

Second, many species killed by pesticides may be non-target organisms. The third problem is persistence; some pesticides have great longevity in the environment, so that non-target effects persist long usage has ceased. The effects can manifest at multiple levels of biological organization; molecules, tissues, organs, individuals of a species, populations of species and communities and ecosystems (Relyea 2009). The magnitude of these effects is determined by; [1] toxicity/potency, [2] exposure time, [3] dose rate, [4] persistence in the environment and [5] bioavailability.

CLASSIFICATION OF PESTICIDES

Pesticides can be grouped or classified according to some common characteristics:

(1) The pests or group of target organisms which they are designed to control (Table 2).

(2) Their chemical structure: Pesticides are often classified according to their chemical family or chemical structure, for example: organophosphate, organochlorine, carbamates and triazines. Pesticides with similar structures show similar characteristics and usually have a similar mode of action. The active ingredients may be either inorganic (e.g. copper sulphate, ferrous sulphate, copper and sulphur) or organic compounds (e.g. captan, pyrethrin, and glyphosate)

(3) How they work: Various classifications exist based on how pesticides work. For example, contact pesticides kill pest as a result of direct contact with the pesticide, whereas systemic pesticides are absorbed and moved to untreated tissues

(4) Their mode of action (site of action) or the way they destroy or control the target pest. Organophosphate and carbamate compounds inactivate the enzyme acetylcholinesterase which is essential for nerve function. Organochlorine compounds disrupt the sodium/potassium balance of nerve fibers, forcing the nerve to transmit continuously while triazines interfere with photosynthesis (Adams 1995).

The release of a large quantity of a potent pesticide often causes high mortalities in various species. However, under normal conditions, most organisms may be exposed to low levels of pesticides, which results in sublethal responses and adversely affect the structure and function of ecosystems. This dose-dependent effect can also be used as a means of classifying pesticides:

Acute toxicity: effects occur immediately following a single exposure to a high concentration of pesticide. It is defined as the amount of a pesticide required to kill 50 percent of a test population and is expressed as an LD50 (median lethal dose) or LC50 (median lethal concentration). The lower the LD50/LC50 value the more toxic the pesticide.

Sublethal effects: refers to effects resulting from a pesticide dosage that is insufficient to kill but can result in adverse effects. These effects may include reduced productivity, inability to feed, behavioral changes or physiological changes.

Chronic toxicity: Chronic toxicity refers to harmful effects produced by long-term exposure to pesticides. Significantly less is known about the chronic toxicity of pesticides than their acute toxicity. Chronic exposure is often associated with carcinogenesis, teratogenesis, mutagenesis, blood disorders (hemotoxic effects), tumors, liver damage, endocrine disruption and reproductive toxicity.

Endocrine disruption: refers to the hormone-altering ability of pesticides acting on the endocrine system. These substances alter the messages sent through hormones resulting in permanent functional changes in the organisms.

Ecosystem changes: Many pesticides are non-biodegradable, and can bioaccumulate in biota. They can cause acute and chronic ecological changes resulting from direct toxicity and loss of species, or by indirect changes such as elimination of predators or keystone species.

Table 2. Pesticides and their potential target groups

Type of Pesticide	Target Pest Group
Disinfectants and sanitizers	Kill or inactivate disease-producing microorganisms (bacteria, viruses, etc.) on inanimate objects
Algicides or Algaecides	Algae
Avicides	Birds
Bactericides	Bacteria
Fungicides	Fungi and Oomycetes
Insecticides	Insects
Miticides or Acaricides	Mites
Molluscicides	Snails
Nematicides	Nematodes
Rodenticides	Rodents
Virucides	Viruses
Herbicides	Kill weeds and other plants that grow where they are not wanted

ACUTE TOXICITY

The acute toxic effect of a pesticide is a measure of its capacity or ability to cause injury, illness or mortality to target organisms exposed to a single dosage over a short duration. It is this unique ability to selectively kill organism that makes them such valuable chemicals. However, they are not always selective but indiscriminately kills both target and non-target organisms. The actual extent of toxicity is dependent on a number of factors such as species sensitivity, exposure concentration, potency and duration of exposure. Edwards et al. (1987) similarly attributed the high toxicity of pyrethroids to a combination of three factors: a sensitive central nervous system (mammal), rather slow hydrolytic detoxification, and the route of exposure.

Acute toxicity is often expressed as the amount or concentration of a pesticide - or its Active Ingredients (AI) - required to kill 50 percent of a target population. It is usually reported as the LD50 (median lethal dose) or LC50 (median lethal concentration) value. For example, the 24-hour LC50 of the insecticide permethrin to rainbow trout is 12.5 ppb which means that 50% of the test species would be killed at this concentration. The lower the LD50 or LC50 values the greater the toxicity of the pesticide to the target organisms. For example, Başer (2003) reports that the 48 hr LC50 of permethrin to *Poecilia reticulata* was 245.7 μg/L while for *Gambussia affinis* it was 2.7 μg/L indicating a higher toxicity to this species. Numerous studies have also reported on the toxicity of pesticides to a wide range of organisms (Table 3). Based on the acute LD50 or LC50 pesticides can be fitted into 1 of 5 toxic categories: class I (no acute toxicity, $TU < 0.4$), class II (slightly acute toxicity, $0.4 < TU < 1$), class III (acute toxicity, $1 < TU < 10$), class IV (high acute toxicity, $10 < TU < 100$), class V (very high acute toxicity, $TU \geq 100$) where TU represents the acute toxic units (Persoone et al. 2003).

Table 3. Toxicity of various pesticides to test organisms

Chemical and test species	Duration (hr)	Toxicity (LC50)	Reference
Carbaryl			
Clarius batrachus	96	5.248 ppm	Wasu et al. (2009)
Clarius garipinius	96	0.38 ppm	Omitoyin et al. (2006).
Malathion			
Clarius batrachus	96	0.25 ppm	Wasu et al. (2009)
Uryzias latipes	48	1.8 mg/L	Tsuda et al. (1997)
Deltamethrin			
Homarus americanus	96	3.74 to 4.92 ng/L	Fairchild et al. (2010)
Eohaustorius estuarius	96	1.66 to 7.99 ng/L	Fairchild et al. (2010)
Crangon septemspinosa	96	27.4to 45.3 ng/L	Fairchild et al. (2010)
Oreochromis niloticus	48	1.7 µg/L	Karasu Benli et al. (2009)
Cyprinus carpio	96	1.65 µg/L	Calta and Ural (2004)
Oncorhynchus mykiss	96	0.6961 µg/L	Ural and Sağlam (2005)
Poecilia reticulata	48	5.13 µg/L.	Viran et al. (2003)
Salmo gairdneri	96	0.39 µg/l	Mestres and Mestres (1992)
Cyprinus carpio	96	1.84 µg/L	Mestres and Mestres (1992)
Sarotherodon mossambica	96	3.50 µg/L	Mestres and Mestres (1992)
Imidacloprid			
Rana limnocharis	96	82 mg/L	Feng et al. (2004)
Rana nigronaculata Hallowell	96	129 mg/L	Feng et al. (2004)
Endosulfan			
Bufo regularis	96	0.73 mg/L	Ezemonye and Tongo (2010)
Diazinon			
Bufo regularis	96	0.44 mg/L	Ezemonye and Tongo (2010)
Puntius gonionotus	96	3.67 mg/L	Hoque et al. (1993)
Labio calbasu	96	1.54 mg/L	Al-Arabi et al. (1992)
Uryzias latipes	48	4.4 mg/L	Tsuda et al. (1997)
Cyprinus carpio	96	9.76 mg/L	Ahmad (2011)
Permethrin			
Cyprinus carpio	48	56.90 mg/L	Davoodi and Abdi (2012)
Poecilia reticulata	48	245.7 µg/L	Başer (2003)
Salmo salar	96	12.0 µg/L	Mc Leese et al. (1980)
Salmo gairdneri	96	43 µg/L	Kumaraguru and Beamish (1981)
Gambussia affinis	48	2.7 µg/L	Milam et al. (2000)
Salvelinus fautinalis	96	3.2 µg/L	Mayer and Ellersieck (1986)
Lindane			
Clarias gariepinus	96	1.29ppm	Lawson et al. (2011)

Organisms often show variations in their acute toxic responses to pesticides. For example, penaeid shrimps are reported to be 36 times more sensitive to organochlorine pesticides and 5 times more sensitive to organophosphate pesticides than mysids. However, mysids were reported to be 54 times more sensitive to pyrethroids than penaeids (Cripe 1994). This variability in acute toxic responses may be related to a number of factors such as: chemical characteristics, age of test species, life stage, body size, health, molt stage, life history, habitat quality, metabolic rates and genetic predisposition. Some pesticides are also more toxic than others, for example, deltamethrin is reported to be two times more toxic to fish than dieldrin. The differences in acute toxicity between species and between chemicals can be attributed to differences in susceptibility and tolerances related to differential absorption rates, their accumulation, biotransformation and excretion of pesticides. For example, Omitoyin et al. (2006) suggested that these were the reasons for the differential toxicity of lindane to *C. gariepinus.*

The acute toxicity of pesticides is also dependent on their stereochemical structure. Edwards et al. (1987) reported that cis- and trans-cypermethrin at concentrations of 0.25 and 0.17 μg/g, respectively, were associated with toxicity in trout. Similarly, alpha-cypermethrin which is a potent neurotoxic, has been shown to be highly toxic to aquatic invertebrates and fishes. The 48 hr EC50 value for *Daphnia magna* is reported to be 0.3 μg/L, while the 96 hr LC50 values for fish species range between 0.7 and 350 μg/L (URL 1). However, alpha-cypermethrin is practically non toxic to birds, mollusk and some higher vertebrates. This is usually common with synthetic pyrethroids, which often demonstrates a species selective toxicity in the order: fish/invertebrates >> mammals >> birds. Pyrethroids have been shown to be lethal to fishes at concentrations 10–1000 times lower than the corresponding values for mammals and birds. This differential toxicity between species may be due to the significantly lower rates of metabolism and elimination and the increased sensitivity of the nervous system in fishes when compared to birds and mammals. The rate of metabolic transformations and the nature of the metabolites depend mainly on the organism involved; for example, birds and mammals metabolize and excrete more rapidly than insects and fishes. Edwards et al. (1987) found that cypermethrin was metabolized and eliminated significantly more slowly by fishes when compared to mammals or birds, which may explain the higher toxicity to fishes. In mice and rats the main reaction involved in the metabolism of the pyrethroids; deltamethrin, cypermethrin or cyhalothrin, is ester cleavage mainly due to the action of carboxyesterase. However, metabolism of pyrethroids in fish is largely oxidative (Demoute 1989) as they

SUBLETHAL EFFECTS

Mortality is not the only endpoint associated with pesticide exposure. More often organisms are exposed to "sublethal" concentrations which may lead to sublethal responses such as changes in: behavior, weight loss, inability to avoid predators, lowered tolerance to stress, reproductive inhibition, growth retardation, life span, development rates, fecundity, egg viability, sex ratio, oxygen consumption, enzyme function and physiological and biochemical functioning. Some pesticides also act as endocrine disruptors, impairing fertility and the development of animals or may bind to sex hormone receptors leading to responses similar to endogenous oestrogens and androgens. They may alter the concentration of hormone receptors in target organs, interfere with the biosynthesis of hormones in the endocrine system, or affect biotransformation processes in the liver. Compounds such as organophosphate, dithiocarbamates, copper fungicides and pyrethroids have been shown to impact spermatogenesis and the number and quality of sperms.

Behavioral Responses

Abnormal behavioral responses, such as reduced activity, loss of equilibrium, vertical positioning, rapid gill movement, erratic swimming, swimming at the water surface, air gulping from the water surface, or staying motionless have been observed in young mirror carp (*C. carpio*), guppies (*P. reticulata*) and freshwater catfish (*Heteropneustes fossilis*) following exposure to sublethal concentrations of pyrethroids, deltamethrin and b-cypermethrin. Karasu Benli et al. (2009) reported that tilapia fry typically showed changes in behavioral response within 30 minutes after exposure to 2.5 µg/L of deltamethrin with loss of balance and erratic swimming occurring within 2 hr of exposure. At 3.0 µg/L they were observed turning on their vertical axis trying to "gulp" air at the surface of the water.

Abnormal behavioral responses can been linked to the inhibition of acetylcholinesterase (AChE) activity. AChE is responsible for the removal of the neurotransmitter acetylcholine (ACh) from the synaptic cleft through hydrolysis into choline and acetic acid (Habig and DiGiulio 1991). ACh activity is important for normal behavior and muscular functions and is therefore a major target site for many toxicants. Inhibition of AChE activity results in the accumulation of ACh in the synaptic cleft, which is a sign of cholinergic toxicity and may lead to hyperstimulation and cessation of neuronal transmission (paralysis) (Halappa and David 2009). Generally, organisms poisoned with anticholinesterase insecticides show signs of muscle paralysis, especially of the fins and respiratory apparatus, hyperactivity and loss of balance (Sancho et al. 1998). Chlorpyrifos for example has been shown to cause irregular, erratic and darting swimming movements, hyper excitability, disrupted schooling behavior, loss of equilibrium and morphological deformities in *Cyprinus carpio* (Halappa and David 2009). Chlorpyrifos undergoes biotransformation through an initial desulfuration reaction involving cytochrome P450 (CYP) followed by dearylation involving cytochrome P450 and A-esterase (Amitai et al. 1998). The biotransformation process results in the formation of an active oxygen analog (chlorpyrifos-oxon) which is known to inhibit acetylcholinesterase (AChE) activity.

Histological and Morphological Responses

Histological and morphological changes have also been observed in various species following exposure to pesticides. Organophosphate pesticides such as methyl parathion have been shown to cause epithelial hyperplasia, epithelial hypertrophy, edema, and detachment in the gills in *Corydoras paleatus.* Desquamation and necrosis of gill epithelia of *Gambusia affinis* following exposure to deltamethrin has also been reported. Lifting of the lamellar epithelium, oedema, and dilatation of the capillaries in primary lamellae, aneurism, epithelial hyperplasia, epithelial necrosis and fusion of the secondary lamellae were also reported (Altinok and Capkin 2007). Lipoid vacuolization of enterocytes, apical cytoplasm, and an increase in goblet cell activity in the intestine have also been reported following ingestion of pesticides. Cloudy swelling, bile stagnation, focal necrosis, atrophy, and vacuolization have also been reported in the liver of various organisms. Hepatic lesions in the liver tissues of fish exposed to deltamethrin characterized by hypertrophy of hepatocytes, significant increase of kupffer cells, circulatory disturbances, focal necrosis, fatty degeneration, nuclear pycnosis and narrowing of sinusoids are also common effects. Infiltration of mononuclear leucocyte and eosinophils towards lamina propria, necrosis were detected in gut (Cengiz and Ünlü 2006; Yildirim et al. 2006). Metabolic processes that depend on liver function are often equally impaired, but secondary effects vary with gill and intestine pathologies (Fanta et al. 2003).

Population and Ecosystem Level Responses

The effects of pesticides can be realized at varying levels: impacting guilds of organisms with similar feeding strategies or resource use; affecting entire taxonomic groups; impacting community structure and ecosystem function. Liess and Schulz (1999) showed that the population density of instream organisms decreased significantly shortly after the occurrence of three highly contaminated runoff events. Eight of the eleven abundant species disappeared, and the numbers of the remaining three species were significantly reduced. In contrast, no negative effect on the macroinvertebrates was detected in the stream during runoff events in which contaminant levels were low. In recent times, pesticides have been linked to major declines in populations of bats, birds, amphibians, invertebrates and other aquatic and terrestrial fauna. An extensive review of the impacts of pesticides on ecosystems has been presented by Pimentel and Edwards (1982) and Yudav (2010).

Adverse ecological effects may occur as a result of; loss in productivity, changes in growth rates, developmental abnormalities, altered diversity, changes in community structure, changes in system processes (such as nutrient cycling) and losses of valuable species. Generally, pesticides could reduce species diversity and contribute to population decline in animals and plants by destroying habitats, reducing food supplies and impairing reproduction. They may eliminate a species essential to the functioning of the entire community, or it may promote the dominance of undesired species or simply decrease the number and variety of species within the ecosystem. Roundup, for example has been shown to cause a 70% decline in the species richness of tadpoles while other pesticides such as Sevin and malathion have been shown to reduce over all species richness by 15% and 30%, respectively (Relyea 2005).

Population level can be affected by a number of different factors, the most important of which is reproductive success. Moore and Waring (2001) reported that low levels of cypermethrin can cause disruption of reproductive function and reduction in the level of fertilization in Atlantic salmon resulting in significant long-term effects on populations and community structure. In salmon, ovulating females release a priming pheromone (F-type prostaglandin, PGF2α) which is detected by male and stimulates an increase in the levels of plasma sex steroids. When males are exposed to < 0.004 mg/L of cypermethrin, it reduces or inhibits the olfactory response to PGF2α, causing a significant reduction in their endocrine response to the priming pheromones. The olfactory system is a vital part of spawning synchronization as it mediates both behavioral and physiological/endocrine responses (Stacey and Sorensen, 2002). In addition, exposure of salmon milt and eggs to 0.1 mg/L of cypermethrin during fertilization can cause a reduction in the number of eggs fertilized. These responses can have a significant impact on populations by reducing overall species numbers. A decrease in the number of fertilized eggs was found when sperm were activated with water containing different concentrations of cypermethrin (Moore and Waring 2001). Jaensson et al (2007) also reported that cypermethrin concentrations of 1 μg/L adversely affected the spawning behavior of brown trout (*Salmo trutta)* which showed a lower frequency of courting behavior and spent less time near the female. Males also had significantly lower *11-ketotestosterone* (11-KT) concentrations and lower amounts of expressible milt in their plasma. Pesticides have also been reported to reduce reproductive success by causing thin eggshells and reducing hormone levels necessary for egg laying in different birds species (Ospreys, Peregrine Falcons, eagles, quills) (Fry et al. 1981). Diazinon and DDT have been linked to widespread mortalities in bird species and have been known to cause; reduce egg numbers, decrease survival of eggs and nestlings, and increase the number of deformities in developing chicks, reduction in the amount of food consumed, weight reduction and changes in physical activity. DDT has also been shown to cause alterations in the sex ratio of some species such as *California gull.* Within these populations DDT exposure has also been shown to cause feminization of embryos resulting in a population showing a skewed sex ratios and reduced numbers of breeding males (Fry et al. 1981).

Pesticides may influence populations in a less direct manner, by reducing the effective population size, that is, reducing the number of individuals that contribute genetic material to the next generation. The net effect is a loss of some alleles and overall decrease in genetic variability. Since genetic variations seed evolutionary changes, it can result in a population with a reduced ability to adapt to future changes in pesticide exposure. However, it means that the new population would become genetically predisposed to be resistant to the pesticide. Multiple resistance - resistance to more than one pesticide and to pesticides in more than one chemical class - is increasing rapidly.

Changes in community structure and ecosystem function generally stems from population level changes. These can arise because:

- The population level of a particular species was lowered sufficiently as a result of reduced growth or reproduction.
- Indirect loss of species because an important keystone species was eliminated directly by the pesticide.

Indirect effects through trophic cascades in communities are common and may have a similar or greater influence on species abundance than direct effects (Lampert et al. 1989; Fleeger et al. 2003). For example, the most frequently reported community response to pesticides is a change from dominance of large zooplankton species to dominance by small zooplankters through relieved competition (Havens and Hanazato 1993; Hanazato 1998). Medina et al. (2004) reported that zooplankton density in mesocosms recovered after pesticides were removed from their environment, but the community structure was altered as copepods were killed and rotifers became the dominant species. Malathion, for example, has been shown to cause a dramatic reduction in zooplankton leading to an alteration of aquatic communities. At intermediate and high concentrations, both cladocerans and copepods were eliminated but at low concentration, the zooplankton assemblage rebounded but was converted from being dominated by cladocerans to being dominated by copepods (Relyea and Hoverman 2008). Hanazato and Yasuno (1987) also suggested that higher concentrations of carbaryl (0.5 to 6.0 mg/L) dramatically reduce all groups of zooplankton whereas lower concentrations (0.1 mg/L) kill the cladocerans but not the copepods.

Introduction of pesticides into the environment can also result in the elimination of some sensitive species such as fishes, which are reportedly 1000 time more sensitive to the effects of pesticides than birds and mammals. Elimination of some species may eventually lead to proliferation of tolerant individuals, or faster growing tolerant species. Hanazato and Yasuno (1987) that carbaryl pesticides for example, caused a reduction in the density of Cladocera in artificial lake systems. This reduction allowed predators such as Chaoborus larvae to prevent recovery of the cladocerans population enabling rotifers to increase and dominate in the zooplankton community.

Occasionally, pesticides may not eliminate a species, but reduce its abundance, intrinsic growth and reproductive potential. It may also selectively eliminate susceptible age groups in a population. Pesticides often affect population growth rates by reducing age-specific survival probabilities (Sibly 1996) and consumption (Baird et al. 1990) which increases generation time and reduces offspring production. Reproduction rates could also be affected as resources and energy are diverted from reproduction to repair or detoxification mechanisms (Baird et al. 1990). An animal's energy budget can be expressed by a simple equation: Pn = Pg - R, where gross and net productivity are represented by Pg and Pn, respectively, and respiration R (Hickman et al. 2006). Energy spent on stress responses increases the demands on respiration (R) ultimately lowing net productivity which impacts the overall fitness of organisms as maintenance must be first satisfied before energy is made available for other processes such as growth and reproduction. Swanepoel et al. (1999) reported that exposure to lindane caused an increase in the metabolic rate of bats which could be compensated for by increased food intake. However, for free living bats, increased food intake also leads to increased foraging activity and time, which increases the risk of predation or accident which would affect their longevity. Halappa and David (2009) reported that fish exposed to chlorpyrifos showed a drastic reduction in food consumption which could be a defense mechanism as it may prove more profitable to decrease food uptake under toxic environmental conditions to lower the energetic costs of digestion. Depression in appetite is a common response of fish to stress and intermittence of feeding for longer periods does have a clear impact on growth, reproduction and survival (Rice 1990). Dembélé et al. (2000) indicated that the abnormalities in fish behavior observed following exposure to insecticides (chlorfenvinphos, chlorpyrifos and diazinon) could be related to failure of energy production or the release of stored metabolic energy, which may cause severe stress, leading to the death. In uncontaminated lakes containing abundant daphnia, a large proportion of phytoplankton primary production is transferred to top predators (fish and birds) through the algae → zooplankton → fish pathway. However, in lakes impacted by insecticides, where there were many small zooplankton and few Daphnia a smaller proportion of primary production is transferred upwards, due to longer food chains that include invertebrate predators. Because energy loss occurs during transfer from one trophic level to another, efficiency of energy transfer from primary producers to top predator is lower in a pesticide-stressed ecosystem than in an ecosystem uncontaminated by chemicals (Hanazato 2001). In prairie grasslands treated with diazinon the reproductive capacity of *Sigmodon hispidus* (cotton rat) which fed on contaminated arthropods was reduced by 33-100% (Sheffield and Lochmiller 2001). Accumulation of dieldrin in British sparrowhawks (*Accipiter nisus*) during the 1960-70s resulted in population declines of up to 60% per year due to a combination of direct mortality and hatching failure (Sibly et al. 2000). Lockhart et al. (1985) also showed that malathion sprayed at 210 g/ha for mosquito control in Winnipeg caused small temporary decreases in both catch of young walleye (*Stizostedion vitreum*) per unit effort and weight gains after the first spray.

Pesticide can also affect swimming behavior which consequently alters predator-prey relationships. They can decrease filtration rates, reduce grazing pressure by zooplankton on phytoplankton altering species abundances and community composition (Ferrando and Andreu 1993, Bengtsson et al. 2004). Relyea (2005) reported that in aquatic communities' species abundance was reduced by 15% with Sevin, 30% with malathion, and 22% with Roundup. Sevin and malathion reduced zooplankton diversity by eliminating cladocerans but not copepods and also reduced the diversity and biomass of predatory insects. Roundup completely eliminated two species of tadpoles resulting in a 70% decline in the species richness of tadpoles.

Pesticides can therefore alter the structure (species richness, density, and biological diversity) and functional activities of an ecosystem. At the community and ecosystem levels pesticides can; [1] alter species dominance, [2] impact species richness and diversity, [3] reduce energy transfer efficiency from primary producers to top predators, and [4] affect reproductive success (Hanazato (2001). The complex food web interactions (indirect effects) within any ecosystem therefore make the assessment of direct pesticide effects a complex process (Preston and Snell 2001).

REFERENCES

Adams, R. W. 1995. Handbook for Pesticide Applicators and Dispensers. Fifth Edition, Victoria, B.C. BC Environment. 248p.

Ahmad, Z. (2011). Acute toxicity and hematological changes in common carp (*Cyprinus carpio*) caused by diazinon exposure. African Journal of Biotechnology. 10(63): 13852-13859.

Al- Arabi, S.S.M., Mazid, M.A., Alam, M.G.A. (1992). Toxicity of Diazinon to three species of Indian major carps. Bangladesh Journal of Training and Development. 5(1): 77-86.

Altinok, İ., Capkin, E. (2007) Histopathology of Rainbow Trout Exposed to Sublethal Concentrations of Methiocarb or Endosulfan. Toxicologic Pathology. 35: 405-410.

Amitai, G., Moorad, D., Adani, R., Doctor, B.P. (1998). Inhibition of acetylcholinesterase and butyrylcholinesterase by chlorpyrifos-oxon. Journal of Biochemical Pharmacology. 56: 293-299.

Başer, S., Erkoç F., Selvi, M., Koçak, O. (2003). Investigation of acute toxicity of permethrin on guppies Poecilia reticulata. Chemosphere. 51; 469–474.

Baird, D.J., Barber, I., Calow, P. (1990). Clonal variation in general responses of *Daphnia magna* Straus to toxic stress. I. Chronic life history effects. Functional Ecology. 4: 399-407.

Bengtsson, G., Hansson, L., Montenegro, K. (2004). Reduced grazing rates in *Daphnia pulex* caused by contaminants: Implications for trophic cascades. Environmental Toxicology and Chemistry. 23(11): 2641-2648.

Calta, M., Ural, M.S. (2004). Acute toxicity of the synthetic pyrethroid deltamethrin to young mirror carp, *Cyprinus carpio*. Fresenius Environmental Bulletin. 13(11a): 1179-1183.

Cengiz, E.I., Ünlü, E. (2006). Sublethal effects of commercial deltamethrin on the structure of the gill, liver and gill tissues of mosquito fish: A microscopic study. Environmental Toxicology and Pharmacology. 21(3), 246- 253.

Cripe, G. M. (1994), Comparative acute toxicities of several pesticides and metals to *Mysidopsis bahia* and postlarval *Penaeus duorarum*. Environmental Toxicology and Chemistry. 13: 1867–1872.

Davoodi, R., Abdi, G. (2012) Comparative Study on the Acute Toxicity of Synthetic Pesticides, Permethrin 25% and Monocrotophos 36%, and Neem-Based Pesticide, Neem Gold EC 0.03%, to Juvenile *Cyprinus carpio* Linn. *Journal* of Biological and *Environmental* Sciences. 6(16): 105-108.

Demoute, J.P. (1989). A brief review of the environmental fate and metabolism of pyrethroids. Pesticide Science. 27: 375–385.

Dembélé, K., Haubruge, E., Gaspar, C. (2000). Concentration Effects of Selected Insecticides on Brain Acetylcholinesterase in the Common Carp (*Cyprinus carpio* L.). Ecotoxicology and Environmental Safety. 45: 49-54.

Edwards, R., Millburn, P., Hutson, D.H. (1987). The toxicity and metabolism of the pyrethroids cis- and trans-cypermethrin in rainbow trout, *Salmo gairdneri*. Xenobiotica. 17(10): 1175–1193.

Ezemonye, L., Tongo, I. (2010). Acute toxic effects of Endosulfan and Diazinon pesticides on adult amphibians (*Bufo regularis*). Journal of Environmental Chemistry and Ecotoxicology. 2(5): 73-78.

Fairchild, W., Doe, K., Jackman, P., Arsenault, J., Aubé, J., Losier, M., Cook, A. (2010). Acute and chronic toxicity of two formulations of the pyrethroid pesticide deltamethrin to an amphipod, sand shrimp and lobster larvae. Canadian Technical Report of Fisheries and Aquatic Sciences. 2876: 34 p.

Fanta, E., Rios, F.S.A., Romao, S., Vianna, A.C.C., Freiberger, S. (2003). Histopathology of the fish *Corydoras paleatus* contaminated with sublethal levels of organophosphorus in water and food. Ecotoxicology and Enviornmental Safety. 54: 119-130

FAO (2009) FAOSTAT On-line Statistical Service. Food and Agriculture Organization of the United Nations.

http://faostat.fao.org/site/424/DesktopDefault.aspx?PageID=424#ancor.

Feng, S., Kong, Z., Wang, X., Zhao, L., Peng, P. (2004) Acute toxicity and genotoxicity of two novel pesticides on amphibian, *Rana N. Hallowell*. Chemosphere. 56(5): 457-63.

Ferrando, M. D., Andreu, E. (1993). Feeding behavior as an index of copper stress in *Daphnia magna* and *Brachionus calyciflorus*. Comparative Biochemistry and Physiology. 106C: 327-331.

Fleeger, J.W., Carman, K.R., Nisbet, R.M. (2003). Indirect effects of contaminants in aquatic ecosystems. Science of the Total Enviornment. 317: 207–233.

Fry, D.M., Toone, C. K. (1981). DDT-induced feminization of gull embryos. Science. 213: 922-924.

Habig, C., DiGiulio, R.D. (1991). Biochemical characteristics of cholinesterases in aquatic organisms. In: P. Mineau, (Ed.), Cholinesterase Inhibiting Insecticides: Their Impact on Wildlife and the Environment - Chemicals in Agriculture, Elsevier, New York, 2: 19-34.

Halappa, R., David, M. (2009). Behavioral responses of the freshwater fish, *Cyprinus carpio* (Linnaeus) following sublethal exposure to chlorpyrifos. Turkish Journal of Fisheries and Aquatic Sciences. 9: 233-238.

Hanazato, T. (1998). Response of a zooplankton community to insecticide application in experimental ponds: a review and the implications of the effects of chemicals on the structure and functioning of freshwater communities. Environmental Pollution 101:361-373.

Hanazato, T. (2001). Pesticide effects on freshwater zooplankton: an ecological perspective. Environmental Pollution. 112: 1-10.

Hanazato, T. and Yasuno, M. (1987). Effects of a carbamate insecticide, carbaryl, on summer phyto- and zooplankton communities in ponds. Environmental Pollution. 48: 145-159.

Havens, K.E., Hanazato, T. (1993). Zooplankton community responses to chemical stressors: a comparison of results from acidification and pesticide contamination research. Environmental Pollution. 82: 277-288.

Hickman C.P., Roberts L.S., Larson A., I'Anson, H., Eisenhour, D. (2006). Integrated Principles of Zoology McGraw Hill. pp 815-818.

Hoque, M.M., Mirja, M.J.A., Miah, M.S. (1993). Toxicity of Diazinon and Sumithion to *Puntus gonionotus*. Bangladesh Journal of Training and Development. 6(1): 19- 26.

Jaensson, A., Scott, A.P., Moore, A., Kylin, H., Olsen, K.H. (2007). Effects of a pyrethroid pesticide on endocrine responses to female odors and reproductive behavior in male parr of brown trout (*Salmo trutta*). Aquatic Toxicology 81: 1-9.

Karasu Benli, A. C., Selvi, M., Sarikaya, R., Erkoç, F., Koçak O. (2009). Acute toxicity of deltamethrin on Nile Tilapia (*Oreochromis niloticus* L.1758) larvae and fry. Gazi University Journal of Science. 22: 1–4.

Kumaraguru, A.K., Beamish, F.W.H. (1986). Effect of permethrin (NRDC- 143) on the bioenergetics of rainbow trout, *Salmo gairdneri*. Aquatic Toxicology. 9: 47-58.

Lampert, W., Fleckner, W., Pott, E., Schober, U., Storkel, K.U. (1989). Herbicide effects on planktonic systems of different complexity. Hydrobiologia. 188/189: 415 –424.

Lawson, E.O., Ndimele, P.E., Jimoh, A.A., Whenu, O.O. (2011). Acute Toxicity of Lindane (Gamma Hexachloro-Cyclohexane) to African Catfish (*Clarias gariepinus*, Burchell, 1822). International Journal of Animal and Veterinary Advances. 3(2): 63-68.

Liess, M., Schulz, R. (1999). Linking insecticide contamination and population response in an agricultural stream. Environmental Toxicology and Chemistry. 18(9):1948-1955.

Lockhart, W.L., Metner, D.A., Ward, F.J. and Swanson, G.M. (1985). Population and cholinesterase responses in fish exposed to malathion sprays. Pesticide Biochemistry and Physiology. 24(1):12-18.

Mayer, F.L., Ellersieck, M.R. (1986) "Manual of Acute Toxicity: Interpretation and Data Base for 410 Chemicals and 66 Species of Freshwater Animals", U.S. Department of the Interior, Fish and Wildlife Service, Resource Publication 160, Washington, D.C., 506p

McLeese, D.W., Metcalfe, C.D., Zitko, V. (1980). Lethality of Permethrin, Cypermethrin, and Fenvalerate to Salmon, Lobster, and Shrimp. Bulletin of Environmental Contamination and Toxicology. 25: 950-955.

Medina, M., Barata, C., Telfer, T., Donald, D.J., Baird, J. (2004). Effects of cypermethrin on marine plankton communities: a simulated field study using mesocosms. Ecotoxicology and Environmental Safety. 58(2): 236–245.

Mestres, R., Mestres, G. (1992). Deltamethrin: uses and environmental safety. Reviews of Environmental Contamination and Toxicology. 124: 1–18.

Milam, C.D., Farris, J.L., Wilhide, J.D. (2000). Evaluating mosquito control pesticides for effect on target and non-target organisms. Archives of Environmental Contamination and Toxicology. 39: 324– 328.

Moore, A., Waring, C.P. (2001). The effects of a synthetic pyrethroid pesticide on some aspects of reproduction in Atlantic salmon (*Salmo salar* L.). Aquatic Toxicology. 52: 1–12.

Omitoyin, B.O., Ajani, E.K., Adesina, B.T., Okuagu, C.N.F. (2006). Toxicity of Lindane (Gamma Hexachloro - CycloHexane) to *Clarias gariepinus* (Burchell 1822). World Journal of Zoology. 1(1): 57-63.

Persoone, G., Marsalek, B., Blinova, I., Törökne A., Zarina, D., Manusadzianas, L., Jawecki, G.N., Tofan, L., Stepanova, N., Tothova, L., Kolar, B. (2003) A practical and user-friendly toxicity classification system with microbiotests for natural waters and wastewaters, Wiley Inter Science, 395-402.

Pimentel, D. and Edwards, C.A. (1982). Pesticide and ecosystems. Bioscience. 32(7): 595-600.

Preston, B.L., Snell, T.W. (2001) Direct and indirect effects of sublethal toxicant exposure on population dynamics of freshwater rotifers. A modeling approach. Aquatic Toxicology. 52: 87-90.

Relyea, R.A. (2005). The impact of insecticides and herbicides on the biodiversity and productivity of aquatic communities. Ecological Applications. 15(2): 618–627.

Relyea, R.A. (2009). A cocktail of contaminants: how mixtures of pesticides at low concentrations affect aquatic communities. Oecologia. 159: 363 – 376.

Relyea, R.A. and Hoverman, J.T. (2008). Interactive effects of predators and a pesticide on aquatic communities. Oikos. 117:1647-1658.

Rice, J.A. (1990). Bioenergetics modeling approaches to evaluate stress in fishes. In proceedings of 8th American Fisheries Society Symposium. Bethesda, Maryland: 80-92.

Sancho, E., Ferrando, M.D., Andreu-Moliner, E. (1998). In vivo inhibition of AChE activity in the European eel, *Anguilla anguilla* exposed to technical grade fenitrothion. Comparative Biochemistry and Physiology. 120: 389-395.

Sheffield, S.R., Lochmiller, R.L. (2001). Effects of field exposure to diazinon on small mammals inhabiting a semienclosed prairie grassland ecosystem. I. Ecological and reproductive effects. Environmental Toxicology and Chemistry. 20(2): 284-296.

Sibly, R.M. (1996). Effects of pollutants on individual life histories and population growth rates. In: Newman, M.C., Jagoe, C.H. (Eds.), Ecotoxicology: A Hierarchical Treatment. CRC Press, Boca Raton, FL, pp. 197–223.

Sibly, R.M., Newton, I., Walker, C.H. (2000). Effects of dieldrin on population growth rates of sparrowhawks 1963–1986. Journal of Applies Ecology. 37(3):540-546.

Stacey, N., Sorensen, P. (2002). Hormonal pheromones in fish. In: Pfaff, D., Arnold, A., Etgen, A., Fahrbach, S., Rubin, R. (Eds.), Hormones, Brains and Behavior. Academic Press, San Diego, California. pp. 375–434.

Swanepoel, R.E., Racey, P.A., Shore, R.F., Speakman, J.R. (1999). Energetic effects of sublethal exposure to lindane on pipistrelle bats (*Pipistrellus pipistrellus*) Environmental Pollution. 104: 169-177.

Tsuda, T., Kojima, M., Harada, H., Nakajima, A., Aoki, S. (1997). Acute toxicity, accumulation and excretion of organophosphorous insecticides and their oxidation products in killifish. Chemosphere. 35: 939-949.

Ural, M.Ş., Sağlam, N. (2005). A study on the acute toxicity of pyrethroid deltamethrin on the fry rainbow trout (*Oncorhynchus mykiss,* Walbaum, 1792): Pesticide Biochemistry and Physiology. 83: 124–131.

URL1. (2008). http://www.inchem.org/documents/jecfa/jecmono/ v38je07.htm.

Viran, R., Erkoç, F.Ü., Polat, H., Koçak, O. (2003). Investigation of acute toxicity of deltamethrin on guppies (*Poecilia reticulata*). Ecotoxicology and Environmental Safety. 55: 82-85.

Wasu, Y.H., Gadhikar A. Y.A., Ade, P.A. (2009). Sublethal and chronic effect of carbaryl and malathion on *Clarias batrachus*. Journal of Applied Sciences and Environmental Management. 13: 23-26.

Yadav, S.K. (2010). Pesticide Applications-Threat to Ecosystems. Journal of Human Ecology. 32(1): 37-45.

Yildirim, M. Z., Benlı, A. Ç. K., Selvı, M., Özkul, A., Erkoç, F. and Koçak, O. (2006), Acute toxicity, behavioral changes, and histopathological effects of deltamethrin on tissues (gills, liver, brain, spleen, kidney, muscle, skin) of Nile tilapia (*Oreochromis niloticus* L.) fingerlings. Environmental Toxicology. 21: 614–620.

In: Pesticides: Characteristics, Uses and Health Implications ISBN: 978-1-62081-577-9
Editors: A. Abrego Gomez and E. Lugo de Ortega

Chapter 7

ADVERSE EFFECTS OF PESTICIDES IN HUMAN HEALTH

Irma M. Medina-Díaz, Aurora E. Rojas-García, María L. Robledo-Marenco, Briscia S. Barrón-Vivanco, Manuel I. Girón-Pérez and Carlos A. Romero-Bañuelos
Universidad Autónoma de Nayarit, México

ABSTRACT

The pesticides are one of the chemical compounds most widely used in the world. The pesticide description commonly referred substances that control organisms which destroy plant life and interfere with the food chain, and which act as vectors for disease organisms to man and animals. This generic definition is frequently extended, rather unsatisfactorily and inaccurately, to cover other chemicals used on plants, such as growth regulators. Pesticides can be classified in different ways, such as by their target, chemical nature, physical state, and mode of action. Classification based on the target (insecticides, herbicides, fungicides, and rodenticides) and chemical structure (organophosphates, carbamates, organochlorines, organochlorines, pyrethroids, phenols, morpholines, chloroalkylthiols, organometallics, azoles, bipyridilium, ureas, anilines, chloronitrile) are the most widely known. Due to their widespread use, a large number of poisonings occur every year in occupationally-exposed human populations, and in the general population appear to regard exposure to pesticides from residues in food and, perhaps, water of greatest concern, there are multiple other sources of exposure which can compound with those from residues. These include hand-to-mouth contact from pesticides used within buildings, veterinary medicines used against domestic pets (flea sprays), and contamination of food and working surfaces from the residential use of pesticides (control of insects). Thus, although the oral route is probably the major route of exposure for the general public, the skin and eyes probably are also significant, and inhalation the least through the consumption of contaminated food and drinking water. The types of adverse health effects that have been documented in exposed workers with pesticides as a group have included acute toxic effects (mirroring the mechanism of toxic action), primary irritancy (the skin and eyes), sensitization (mainly allergic contact dermatitis, but on occasion respiratory sensitization as been described), peripheral and central neurotoxic effects, myonecrosis and in epidemiology studies, there have been suggestions of cardiovascular toxicity, reproductive and developmental toxicity, endocrine alterations, genotoxicity, oncogenicity and immunotoxic effects among other. This review article considers the chemistry, characteristics, uses and health effects of representative groups of pesticides.

PESTICIDES AND THEIR CLASSIFICATION

Definition of Pesticides

Pesticides are a group of chemicals used predominantly in agriculture and against disease vectors such as flies carrying malaria. For most people around the world pesticides are considered simply as poisons. Several international organizations concerned with chemistry, use and regulation of pesticides have established a clear and extended definition. The council of the International Union of Pure and Applied Chemistry (IUPAC) defines a pesticide strictly as a substance intended to kill pests; in common usage, any substance used for controlling, preventing, or destroying animal, microbiological or plant pests (Duffus, 1993; IUPAC, 2011). This definition can only be applied with knowledge of the intended use of the substance.

Almost any substance can be a pesticide if it is used for that purpose. For example, acetone can be used to kill most insects, but it is unusual to use it for this purpose and so it is not normally classified as a pesticide. Similarly, sodium chloride can kill most plants, but is not classified as a pesticide (Nordberg *et al.* 2007). According to the Food and Agriculture Organization of the United Nations (FAO, 2001), a pesticide is "any substance or mixture of substances intended for preventing, destroying or controlling any pest, including vectors of human or animal disease, unwanted species of plants or animals causing harm during or otherwise interfering with the production, processing, storage, transport or marketing of food, agricultural commodities, wood and wood products or animal feedstuffs, or substances which may be administered to animals for the control of insects, arachnids or other pests in or on their bodies. The term includes substances intended for use as a plant growth regulator, defoliant, desiccant or agent for thinning fruit or preventing the premature fall of fruit, and substances applied to crops either before or after harvest to protect the commodity from deterioration during storage and transport."

History of Pesticides

From the beginning of agriculture approximately 10,000 years ago, pests have been a major problem for crop production. The first recorded use of pesticides is 4500 years ago by the Sumerians, who used sulfur compounds to control insects and mites, while about 3200 years ago, the Chinese were using mercury and arsenical compounds for controlling body lice. Pyrethrum, which is derived from the dried flowers of Chrysanthemum cinerariaefolium ("Pyrethrum daisies"), has been used as an insecticide for over 2000 years. Many organic (plant extracts) and inorganic (salt or sea water) chemicals have been used since ancient times as pesticides; Bordeaux Mixture, comprised of copper sulfate and lime, is still used against various fungal diseases (Fishel, 2009; IUPAC, 2010).

Table 1. Important eras in pesticide development

Period	Example	Source	Characteristics
1800-1920	Early organics, nitro-phenols, chlorophenols, creosote, naphthalene, petroleum oils.	Organic chemistry, byproducts of coal gas production, etc.	Often lack specificity and were toxic to user or nontarget organisms.
1945-1955	Chlorinated organics, DDT, HCH, chlorinated cyclodienes.	Organic synthesis.	Persistent, good selectivity, good agricultural properties, good public health performance, resistance, harmful ecological effects.
1945-1970	Cholinesterase inhibitors, organophosphorus compounds, carbamates.	Organic synthesis, good use of structure-activity relationships.	Lower persistence, some user toxicity, some environmental problems.
1970-1985	Synthetic pyrethroids, avermectins, juvenile hormone mimics, biological pesticides.	Refinement of structure activity relationships, new target systems.	Some lack of selectivity, resistance, costs and variable persistence.

Adapted from Stephenson and Solomon (1993).

In general, Zhang *et al.* (2011) suggested that the history of pesticides can be divided into three phases: during the first phase (before the 1870s), natural pesticides were used to control pests; the second phase was the era of inorganic synthetic pesticides (the 1870s - 1945), during which mainly natural materials and inorganic compounds were used; the third phase (since 1945) is the era of organic synthetic pesticides. Since 1945, man-made organic pesticides, e.g., dichlorodiphenyltri chloroethane (DDT), and later hexchlorocyclohexane (HCH) and dieldrin, have ended the era of inorganic and natural pesticides. The use of modern pesticides in agriculture and for public health dates back to the 19th century (Table 1).

Pesticides are used extensively throughout the world, with about 3 billion tons of pesticides applied each year. However, despite the large amount of pesticide applied worldwide, pests, insects, weeds and plant pathogens destroy about 40% of all crops (Pimentel, 2009). In addition, there are more than 26 million cases of non-fatal pesticide poisonings (Richter, 2002). Of all the pesticide poisonings, approximately 3 million result in hospitalization, with approximately 220,000 fatalities and 750,000 developing chronic illnesses every year (Hart and Pimentel, 2002).

Classification and Use of Pesticides

Pesticides can be classified in different ways, including by their target, chemical nature, physical state, and mode of action. The most common pesticide classification is based on the target organism and the chemical structure. According to target classification, four groups of pesticides are the most used. Before the 1990s, the insecticides were the most used; after the 1990s, herbicides are the most commonly used pesticides worldwide by far (45%) followed by insecticides (25%), fungicides/bactericides (24%), and all others (3%) (Zhang *et al.* 2011).

In contrast, under chemical classification, pesticides are categorized according to the chemical nature of the active ingredients into four primary groups; organochlorine (OCPs), organophosphate (OPs), carbamate and pyrethroid (Table 2). OCPs were used commonly in the past, but many have been removed from the market due to their persistence as well as their health and environmental effects (e.g. DDT and chlordane). OPs, which were used in World War II as nerve agents, replaced OCPs, and do not persist in the environment.

Carbamates are organic compounds derived from carbamic acid and carbamate esters, also called urethanes. Pyrethroid pesticides were developed as a synthetic version of the naturally occurring pesticide pyrethrin, found in chrysanthemums. They have been modified to increase their stability in the environment. Some synthetic pyrethroids are toxic to the nervous system (EPA, 2011).

Another important criterion for the classification of pesticides is the toxicity level, which is associated with health implications. In this context, the World Health Organization (WHO) has developed a classification system that group pesticides according to the potential risks to human health.

The WHO Recommended Classification of Pesticides by Hazard was approved by the 28th World Health Assembly in 1975 and has gained wide acceptance. Guidelines were first issued in 1978, and have since been revised and reissued every few years.

The classification distinguishes between the more and the less hazardous forms of each pesticide, in that it is based on the toxicity of the technical compound and on its formulations. The classification is based primarily on the acute oral and dermal toxicity to rats (Table 3).

Despite technological advances, it is estimated that 35 - 40% of global agricultural production is lost due to pest activity. This provides a strong incentive to develop methods to mitigate the problems caused by pests (Peshin, 2002).

Table 2. Classification of pesticides based on their chemical structure

Class	Example
Organochlorines	DDD, DDT, DDE, chlordane, kepone, dieldrin, endosulfan, heptachlor, lindane, mirex, methoxychlor, toxaphene
Organophosphorus	chlorpyrifos, glyphosate, diazinon, dimethoate, malathion, methamidophos, parathion, terbufos, tribufos, trichlorfon
Carbamates	aldicarb, carbaryl, carbofuran, fenoxycarb,propoxur
Pyrethroids	cypermethrin, fenvalerate, permethrin, pyrethrin, pyrethrum, resmethrin, tetramethrin
Anilides/anilines	metolachlor, pretilachlor, propachlor, trifluralin
Phenoxy	2,4-D, 2,4-DB, 2,4,5-T, MCPA, MCPB, fenoprop
Triazines	atrazine, cyanazine, hexazinone, prometryn, propazine, simazine, terbutryn quaternary diquat, MPP, paraquat
Ureas	chlortoluron, DCMU, metsulfuron-methyl, monolinuron

Table 3. The WHO recommended classification of pesticides by hazard

Class		LD_{50} for the rat (mg/kg body weight)			
		Oral		Dermal	
		Solids	Liquids	Solids	Liquids
Ia	Extremely hazardous	5 or less	20 or less	10 or less	40 or less
Ib	Highly hazardous	5 - 50	20 - 200	10-100	40 – 400
II	Moderately hazardous	50 - 500	200 - 2000	100-1000	400 – 4000
III	Slightly hazardous	Over 500	Over 2000	Over 1000	Over 4000
U	Unlikely to present acute hazard in normal use				

Adapted from WHO (2004).

Currently, the main use of pesticides continues to be agriculture; it is estimated that agricultural use accounts for 77% of pesticides manufactured in the world. From 1964 to date, pesticide use increased by 170%. The geographic distribution of pesticide use depends on the culture. However, currently the market of these substances is concentrated in three geographical areas: North America, Western Europe and East Asia. It is estimated that, countries in North America use 30% of world pesticide production, while Europe uses 27%, followed by Japan with 12%. Developing countries, including China, used approximately 31% (Smith and Bridges, 1996; Hodge *et al.* 1993).

Non-Agricultural Use of Pesticides

In recent years, the use of pesticides has been a boom for non-agricultural activities such as weeding near roads (highways and railways), cemeteries, recreational parks and golf courses. As well as use for important eliminating pests, such as rodents and flies, for public health reasons. In this context, it is estimated that these activities consume about 550 tons of pesticides per year, or 25% of world pesticide production.

Health Implications

Chlorophenoxy Acids

Information on the human toxicity of chlorophenoxy herbicides derives mostly from occupational medicine. In many studies and reports, workers were acutely or chronically exposed to one or several chemicals at the same time. A few cases of acute poisoning by accidental or voluntary ingestion of pure 2,4-dichlorophenoxyacetic acid (2,4-D) and/or 2,4,5-trichlorophenoxyacetic acid (2,4,5-T) have been described. However, in these cases, the chlorophenoxy acids showed low to moderate acute toxicity and caused local irritation and burning of the skin and mucosae, muscle dysfunction and weakness, hypotension, arrhythmia and respiratory difficulty. An oral lethal dose in humans of more than 6 g has been estimated for 2,4-D (Murphy, 1986).

A number of epidemiological studies have shown a higher tumor-related mortality in subjects exposed to chlorophenoxy pesticides (Axelson *et al.* 1980). A case-control study

(Harden and Sandstrom 1979; Eriksson *et al.* 1981; Hardell *et al.* 1981) suggested a possible association between exposure to chlorophenoxy herbicides and soft tissue sarcomas or non-Hodgkin's lymphoma (NHL). Such an association has been confirmed by some, but not all subsequent studies (Coggon and Acheson, 1982; Pearce *et al.* 1986; Hoar *et al.* 1986; Dalager *et al.* 1991). The case-control and cohort epidemiology studies did not confirm a causal relationship between exposure to these herbicides and cancer (Bond *et al.* 1989; Johnson, 1990).

Triazines

Epidemiological carcinogenicity studies on triazines evaluated by IARC (1991) led to the conclusion that exposure to triazine herbicides may be carcinogenic to humans, although the specific role of individual compounds could not be determined.

As to the acute or chronic toxicity of triazines in humans, no reports of systemic poisoning by atrazine or simazine have been found in the literature. Only occasional episodes of eye irritation have been reported with atrazine.

Bipyridylium Compounds

Reports of human exposure to paraquat resulting in documented toxic effects are almost invariably limited to accidental or voluntary (suicide) ingestion of concentrated solutions of the compound. In a few cases, however, absorption by damaged skin — the compound is poorly absorbed by the intact skin — and subcutaneous or intravenous injection has been reported. The main pathological changes in the pulmonary parenchyma observed in a typical case of paraquat poisoning are the destruction of alveolar epithelial cells, capillary hemorrhage, and pulmonary edema two or three days after ingestion, followed by interstitial fibroblast proliferation, and finally, diffuse fibrosis in about two weeks. Changes in the kidney (tubular degeneration) and in other organs may also be observed, usually at higher doses. Local chronic effects of paraquat have been described after occupational exposure, usually in sprayers or manufacturers. Systemic poisoning may also occur following dermal absorption of paraquat. Several cases have been reviewed (Smith, 1988). These include skin hyperpigmentation and hyperkeratosis, nail damage and eye irritation. Nasal and throat irritation may also result from inhalation exposure to aerosols of paraquat solutions (Smith, 1988).

Organochlorine

Both experimental studies and epidemiological studies indicate that some OCPs such as DDT and dichlorodiphenyldichloroethylene (DDE) are neurodevelopmental toxicants (Eskenazi *et al.* 2006). The central nervous system of a child may be particularly vulnerable to neurotoxicity from environmental agents since it is still developing. Environmental agents such as DDT may easily disturb the process by changing neurotransmitter systems and altering intracellular signaling processes resulting in behavioural impairments (Kodavanti,

2006). Also, several studies suggest that prenatal exposure to OCPs is associated with an increase in attention deficit hyperactivity disorder (ADHD) symptoms among children at 4 years of age (Darvill *et al.* 2000; Jacobson and Jacobson, 2003; Ribas-Fito *et al.* 2007; Sagiv *et al.* 2010). Serum OCPs concentrations measured at ages 8 to 9 years were associated with reduced growth, particularly reduced BMI, during the peri-pubertal period, which may impact attainment of optimal adult body mass and height. Prenatal exposure to p,p'-DDE and hexachlorobenzene (HCB) has been associated with reduced birthweight and length, independent of gestational age (Eggesbo *et al.* 2009; Ribas-Fito *et al.* 2002; Siddiqui *et al.* 2003; Weisskopf *et al.* 2005; Wolff *et al.* 2007).

Epidemiological studies show a consistent association between pesticide exposure and Parkinson's disease (Dutheil *et al.* 2010). In a recent study in France, phenoxy compounds were associated with Parkinson's disease, although this only reached statistical significance among men older than 65 years (Weisskopf *et al.* 2010). Recent epidemiological studies reported that background exposure to OCPs was strongly associated with type 2 diabetes in the general population (Rylander *et al.* 2005; Porta 2006; Montgomery *et al.* 2008). Among several subclasses of OCPs showed the strongest association with type 2 diabetes (Lee *et al.* 2006; Lee *et al.* 2007). DDT, heptachlor, lindane, chlordane, and maneb/mancozeb were also associated with hyperthyroidism (Goldner *et al.* 2010).

Experimental findings suggesting that OCPs exert weak estrogenic, antiestrogenic, and antiandrogenic effects (Bonefeld-Jorgensen *et al.* 2001; Danzo, 1997; Kelce *et al.* 1995) have raised concern over their potential to influence the development of testicular germ cell tumors (TGCTs) and other male reproductive disorders speculated to be part of the testicular dysgenesis syndrome: cryptorchism, hypospadias, and impaired spermatogenesis (Skakkebaek *et al.* 2001; Toppari *et al.* 1996; Purdue *et al.* 2009). p,p'-DDE, oxychlordane, trans-nonachlor compounds, and, more speculatively, some PCB congeners (44 and 167 in particular) may be associated with risk of TGCTs including seminomas (Lee *et al.* 2006). These pesticides have been also strongly associated with type 2 diabetes, insulin resistance, and metabolic syndrome (Lee *et al.* 2007; Lee *et al.* 2008).

Organophosphates

The principal mechanism of acute toxicity of OPs is based on inhibition of the acetylcholinesterase enzyme (AChE), which is responsible for the degradation of the neurotransmitter acetylcholine. Acetylcholine stimulates postsynaptic responses at ganglionic and cholinergic synapses in the peripheral and central nervous system and at neuromuscular junctions. Inhibition of this enzyme causes the continued presence of the neurotransmitter into the synapse, with the corresponding postsynaptic receptor stimulation and retention of electrical activity.

The interaction of AChE with OP occurs in an amino acid serine hydroxyl substituent, with the formation of a phosphorylated enzyme, very stable and virtually inactive. The phosphate group in the moiety of insecticide has an important role in the anticholinesterasic capacity. OP bind to AChE irreversibly, so the nervous system only recovers normal function when it has synthesized enough AChE to replace the inactive enzyme. OP not only reacts irreversibly with AChE, but the phosphorylated enzyme tends to experience "aging" caused by enzyme dealkylation, which strengthens the bond between the phosphate group and the active site of the enzyme, and is a major barrier to enzyme reactivation (Moreno, 2003).

The symptoms associated with acute OP toxicity (Table 4) usually disappear within days or weeks after exposure has ended, but others are persistent and can be present for months or years with neuropsychological symptoms that point to damage the central and peripheral nervous systems.

Table 4. Symptoms associated with acute OP toxicity

Receptor type	System	Organ or system	Effect
Muscarinic	Parasympathetic	Eyes	Miosis and blurred vision
		Gastrointestinal tract	Nausea, vomiting, diarrhea and incontinence
		Exocrine glands	Salivation, and sweating, tearing
		Respiratory	Excessive bronchial secretion, edema, bronchospasm, bronchoconstriction, cough, bradycardia and increased blood pressures
		Urinary bladder	Incontinence
Nicotinic	Parasympathetic and sympathetic	Cardiovascular	Opposed to the muscarinic effects with tachycardia and increased stress
		Motor nerve fibers	Fasciculation, cramps generalized muscle weakness and respiratory muscle peripheral, paralysis, tremors, convulsions, depression of the respiratory centers, cyanosis and coma

There is a significant variation in the ability of OPs to inhibit AChE. The degree of inhibition is related to the lipid-water partition of the OP (reviewed in Ray and Richards, 2001). OPs also inhibit other enzymes such as acid phosphatases, lipases, trypsinases, chymotrypsinases and dehydrogenases, among others (reviewed in Ray and Richards, 2001).

Regarding reproductive effects, studies in men and animals have demonstrated that OP alter male reproductive function, through sperm chromatin condensation, semen quality and hormone balance (Padungton *et al.* 2000; Sánchez-Peña *et al.* 2004; Piña-Guzmán *et al.* 2005; Recio *et al.* 2005). In addition, the oxon OP metabolites can alter sperm DNA integrity, to an even greater degree than their parent compounds (Salazar-Arredondo, 2008). Some studies that evaluated OP genotoxicity in humans report a significant association between occupational exposure to complex mixtures of pesticides and the presence of chromosomal aberrations, sister chromatid exchanges and micronuclei, mainly in farmers. Most of these studies found dose-dependent effects and a direct relationship with time and the degree of exposure. The chromosomal damage caused by pesticide exposure seems to be transient in acute exposure, but chronic exposures result in the accumulation of complex mixtures of chemicals (Bolognesi, 2003; 2010).

Among the potential mechanisms of OP genotoxicity is the induction of oxidative stress. One possible origin of oxidative stress (ROS generation) is through cytochrome P450 activation during OP metabolism, producing mainly superoxide. For example, damage caused by acute methyl parathion exposure is involved in the DNA damage observed in mice epididymal spermatozoa (Piña-Guzmán *et al.* 2006). Quinalfos exposure caused testicular damage due to lipid peroxidation, along with alterations in some antioxidant enzymes and reduced glutathione levels (Debnath and Mandal, 2000). Although the generation of oxidative damage by OP is a very plausible mechanism of their toxicity on sperm DNA, active oxons also exert genotoxic effects probably by an oxidative mechanism (Salazar-Arredondo, 2008).

On the other hand, long-term exposure to low levels of OP as well as other pesticides may produce neuropsychiatric symptoms. Most studies of moderate pesticide exposure have found increased prevalence of neurological symptoms and changes in neurobehavioral performance, reflecting cognitive and psychomotor dysfunction. There is less evidence that moderate exposure is related to deficits in sensory or motor function or peripheral nerve conduction. It is possible that the most sensitive manifestation of pesticide neurotoxicity is a general malaise lacking in specificity and related to mild cognitive dysfunction. Pesticide exposure may also be associated with increased risk of Parkinson's disease; several classes of pesticides, including insecticides, herbicides, and fungicides, have been implicated. Studies of other neurodegenerative diseases are limited and inconclusive (reviewed by Kamel and Hoppin, 2004). Major unresolved issues include the relative importance of acute and chronic exposure, the effect of moderate exposure in the absence of poisoning, and the relationship of pesticide-related neurotoxicity to neurodegenerative disease.

Carbamates

Although the mode of action of carbamates is essentially identical to OPs, there are marked differences in the duration of AChE inhibition. Both OPs and carbamates react with serine hydroxyl groups in the active site of the enzyme. However, the carbamate bond formed is not stable, and the product formed decomposes to regenerate the enzyme, while phosphorylation is virtually irreversible.

The symptoms of carbamate poisoning are almost similar to those described for OPs, but with a lower intensity and duration. In addition to the reversible nature of their connection with AChE, carbamates are rapidly metabolized after absorption. Nevertheless, there are accidents which results in death from acute exposure to high doses of carbamates, especially carbaryl. Also, lasting neurotoxic effects associated with exposure to one massive dose or repeated exposure to high doses have been reported (Moreno, 2003).

Pyrethroids

Pyrethroids are chemicals substances very similar structurally to the pyrethrins, but are generally more toxic to insects and to mammals, because they are combined with synergists that prevent degradation by enzymes (ATSDR, 2003). Their adverse toxic effects are mainly due to persistence in the environment over time, and their consequent interaction with human biological systems, which may in turn lead to a plethora of neuropathological and behavioral health consequences (Shafer and Meyer, 2004). These insecticides are not highly toxic to mammals, but they do accumulate quickly. Although they are sensitive to hydrolysis by nonspecific carboxylesterases, the system of tissue microsomal monooxygenases present in almost all species are extensively involved in pyrethroid bioinactivation in mammals (Ecobichon, 2005; Castillo-Hernandez, 2006).

The pyrethroids are absorbed primarily by the gastrointestinal and respiratory routes, but, in a lesser amount, through the skin. The lipophilic nature of pyrethoids means they are widely distributed, metabolized and excreted rapidly through the feces, breath and urine, mainly in the latter, so do not accumulate (ATSDR, 2003, Castillo-Hernandez, 2006). However, if exposure levels are very high or if exposure is prolonged, pyrethroids can accumulate in fatty tissue and remain in the body for a long time. Certain types of pyrethroids can also be retained for a long time in the skin and hair.

The toxicity of these compounds appears to depend on the rate of hydrolysis (ATSDR, 2003). In humans, the primary toxic effects associated with pyrethrins are derived from their allergenic properties, as they may cause contact dermatitis, pseudo-asthma crisis, anaphylactic reactions and peripheral circulatory failure (ATSDR, 2003; Ecobichon, 2005; Cordova-Palacio, 2006).

Occupational exposure produces variable toxicity. When poured on the head, face or eyes, these agents cause pain, tearing, photophobia, congestion and edema of the conjunctiva and eyelids. Ingestion results in epigastric pain, nausea, vomiting, headache, dizziness, anorexia, fatigue, chest tightness, blurred vision, numbness, palpitations, muscle twitches and disturbances of consciousness. In severe poisonings, seizures with opisthotonus and loss of consciousness have been reported (Ecobichon, 2005). In children, exposure to these compounds cause similar effects, although pyrethroids may penetrate the skin of children more easily than adults (ATSDR, 2003).

Pyrethroids interfere with the normal functioning of nerves and the brain (Anyanwu *et al.* 2004). Certain features of neural tissues make them particularly susceptible to insecticide effects: they are unable to replace lost and damaged cells; some of their components, such as myelin, have the tendency to accumulate and retain lipophilic substances; and they have a large surface area, which increases the likelihood of exposure to systemically distributed neurotoxicants (Desneux *et al.* 2004).

Potentially serious neurotoxic effects of insecticides include peripheral neuropathy, extrapyramidal disorders, ataxia, sensory disorders, psychosis, CNS depression, and in extreme cases, death (Gillette and Bloomquist, 2003). The main acute neurotoxic manifestations suggest a state of hyperexcitability and include convulsions, seizures and paralysis.

The mechanism that produces these effects is the result of esters of type I and II pyrethroids that affect the activation and deactivation of sodium channels. They also inhibit calcium channels and calcium-magnesium ATPase (Kolaczinski and Curtis, 2004; Shafer *et al.* 2005; Ecobichon, 2005). Chronic toxicity generates peripheral distal axonopathy chronically type Wallerian degeneration (Castillo-Hernandez, 2006). There is no evidence that pyrethroids produce birth defects or affect fertility, or any association with cancer (Anyanwu, *et al.* 2004). However, animals chronically exposed to pyrethroids in high amounts develop cancer and adverse reproductive effects (ATSDR, 2003).

Fungicides

Most fungicides have limited toxicity in mammals. However, all these compounds are cytotoxic, most are mutagenic *in vitro*, and almost 90% of agricultural fungicides are carcinogenic in animal models (Ecobichon, 2005). Many fungicides employ copper salts, specifically copper oxychloride and copper sulfate, which are linked to protoplasm coagulation, killing spores and fungi. In humans, these can cause sneezing, nausea, and liver and kidney damage (Vallejo-Rosero, 2006). Others employ organomercury compounds, which attack the central nervous system, producing irreversible damage, such as memory loss, paralysis, blindness, and insanity (Aschner and Ceccatelli, 2010). Another category of fungicides is the dithiocarbamates. These chemicals are environmentally stable, provide good protection and have limited acute toxicity.

Among the most frequently used are maneb, zineb and nabam, singly or in combination, with copper and sulfur salts. These are soluble compounds, not volatile, and not cholinesterase inhibitors. Mancozeb is associated with morphological abnormalities of sperm. Similarly, maneb also has been associated with reproductive toxicity, including a decrease in pregnancy rates, female hormonal disorders and abnormalities of fetal development, as well as damage to testicular function, causing deleterious effects on human reproduction; these lesions may be irreversible (Ecobichon, 2005, Vallejo-Rosero, 2006). Maneb also produces upper respiratory tract irritation, with rhinitis and cough, as well as conjunctivitis and skin outbreaks.

Constant exposure to dithiocarbamate can cause thyroid dysfunction (specifically an enlarged thyroid gland) and also affect the hepatic system, renal system and central nervous system (reflected in tremors and loss of muscle strength). In animals, dithiocarbamate causes personality changes and lung cancer, so it must be handled with extreme caution (Vallejo-Rosero, 2006). The metabolism of these substances in the environment and during the cooking of food, produces ethylenethiourea, which is a carcinogen, mutagen, teratogen (Ecobichon, 2005, Vallejo-Rosero, 2006).

The final common fungicides are the chlorophenols and nitrophenols, which are used as insecticides, molluscicides, acaricides, fungicides and herbicides. Pentachlorophenol dinitroortocresol, dinitrocresol and dinitrophenol are substituted phenolic compounds that are volatile, lipid soluble, and have high acute toxicity. These toxins uncouple oxidative phosphorylation, raising the metabolic rate and body temperature accordingly. Because hyperthermia produces profuse sweating, intoxication may be confused with that caused by OPs. Persons exposed to dinitroortocresol occupationally have yellowing skin, mucous membranes and conjunctiva, and may develop cataracts as a consequence (Vallejo-Rosero, 2006).

Studies from a number of countries and a variety of exposure situations suggest that pesticides may contribute to the development of NHL (Zam and Blair, 1992). A nested case-control study within a cohort mortality study of 22,938 grain millers reported a 4-fold excess of NHL among flour mill employees. Risk rose to 9-fold after 25 years of follow-up (Alavanja *et al.* 1990).

Conclusion

Pesticides are valuable chemicals whose agricultural use is likely to continue and, possibly, increase in the near future. Their use in some cases is associated with a significant toxicological risk for man. A better understanding of the mechanism of toxicity in mammals will improve the toxicological assessment of these chemicals and, together with a more controlled use and a more appropriate perception of the risk, limit the occurrence of new cases of human toxicity.

References

Alavanja M.C.R., Blair A., Masters M.N. (1990). Cancer mortality in the U.S. flour industry. *J. Nati. Cancer Inst.* 82: 840-848.

Anyanwu E.C., Ehiri J.E., Kanu I., Morad M., Ventegodt S., Merrick J. (2004). Assessing the health effects of long-term exposure to insecticide-treated mosquito nets in the control of malaria in endemic regions. *ScientificWorld Journal*. 19: 978-88.

Aschner M., Ceccatelli S. (2010). Are Neuropathological Conditions Relevant to Ethylmercury Exposure? *Neurotox. Res.* 18: 59-68.

ATSDR (2003). *Toxicological profile for pyrethrins and pyrethroids*. http://www.atsdr.cdc.gov/toxprofiles/tp155.pdf [Last revision: January, 2012].

Axelson O., Sundell L., Andersson K., Edling C., Hogstedt C., Kling H. (1980). Herbicide exposure and tumor mortality: an updated epidemiological investigation on Swedish railroad workers. *Scand. J. Work Environ. Health.* 6: 73-79.

Bond G.G., Bodner K.M., Cook R.R. (1989). Phenoxy herbicides and cancer: insufficient epidemiologic evidence for a causal relationship. *Fund. Appl. Toxicol.* 12: 172-188.

Bonefeld-Jorgensen E.C., Andersen H.R., Rasmussen T.H., Vinggaard A.M. (2001). Effect of highly bioaccumulated polychlorinated biphenyl congeners on estrogen and androgen receptor activity. *Toxicology*. 158: 141–153.

Bolognesi C. (2003). Genotoxicity of pesticides: a review of human biomonitoring studies. *Mutation Research*. 543: 251-272.

Bolognesi C., Creus A., Ostrosky-Wegman P., Marcos R. (2011). Micronuclei and pesticide exposure. *Mutagenesis.* 26: 19-26.

Castillo-Hernández L. (2006). Toxicología del sistema nervioso. In: *Toxicología Básica*. Jaramillo-Juárez F., Rincón-Sánchez AR., Posadas del Río FA. Textos Universitarios. 1er Ed. Aguascalientes, México pp.

Coggon D., Acheson E.D. (1982). Do phenoxy herbicides cause cancer in man? *Lancet, i,* 1057-1059.

Córdoba-Palacio D. (2006). *Toxicología*. 5ta Ed. Bogotá. El Manual Moderno. 1044 pp.

Dalager N.A., Kang H.K., Burt V.L., Weatherbee L. (1991). Non-Hodgkin's lymphoma among Vietnam veterans. *J. Occup. Med.* 33: 774-779.

Danzo B.J. (1997). Environmental xenobiotics may disrupt normal endocrine function by interfering with the binding of physiological ligands to steroid receptors and binding proteins. *Environ. Health Perspect.* 105: 294–301.

Darvill T., Lonky E., Reihman J., Stewart P., Pagano J. (2000). Prenatal exposure to PCBs and infant performance on the fagan test of infant intelligence. *Neurotoxicology*. 21: 1029-1038.

Debnath D., Mandal T.K. (2000). Study of quinalphos (an environmental oestrogenic insecticide) formulation (Ekalux 25 E.C.)-induced damage of the testicular tissues and antioxidant defence systems in Sprague–Dawley albino rats. *J. Appl. Toxicol.* 20: 197–204.

Desneux N., Rafalimanana H., Kaiser L. (2004). Dose-response relationship in lethal and behavioural effects of different insecticides on the parasitic wasp Aphidius ervi. *Chemosphere*. 54: 619–627.

Duffus J.H. (1993). Glossary for chemists of terms used in toxicology (IUPAC Recommendations. *Pure Appl. Chem*. 65: 2003-2122.

Dutheil F., Beaune P., Tzourio C., Loriot M.A., Elbaz A. (2010). Interaction between ABCB1 and professional exposure to organochlorine insecticides in Parkinson disease. *Arch. Neurol.* 67: 739-45.

Ecobichon D.J. (2005). Efectos tóxicos de los pesticidas. In: *Casarett y Doull.* Fundamentos de Toxicología. Klaassen CD. Watkins JB. McGraw-Hill Interamericana, pp. 339-353.

Eggesbø M., Stigum H., Longnecker M.P., Polder A., Aldrin M., Basso O., Thomsen C., Skaare J.U., Becher G., Magnus P. (2009). Levels of hexachlorobenzene (HCB) in breast milk in relation to birth weight in a Norwegian cohort. *Environ Res*. 109: 559-66.

Eriksson M., Hardell M., Berg N.O., Moller T., Akelson O. (1981). Soft-tissue sarcomas and exposure to chemical substances: a case-referent study. *Br. J. Ind. Med.* 38: 27-33.

Eskenazi B., Marks A.R., Bradman A. (2006). In utero exposure to dichlorodiphenyltrichloroethane (DDT) and dichlorodiphenyldi chloroethylene (DDE) and neurodevelopment among young Mexican American children. *Pediatrics.* 118: 233-241.

EPA (2011). Environmental Protection Agency. About Pesticides, Types of Pesticides.

FAO (Food and Agriculture Organization of the United Nations). (2001). *5th FAO Consultation on Obsolete, unwanted and banned pesticide stocks Rome,* 10-11 May 2001. Available:http://www.fao.org/fileadmin/user_upload/obsolete_pesticides/docs/consult_5_e.pdf [accessed 8 January 2012].

Fishel F.M. (2009). *Pest Management and Pesticides: A Historical Perspective.* Series of the Agronomy Department, Florida Cooperative Extension Service, Institute of Food and Agricultural Sciences, University of Florida. PI219: 1-6.

Gillette J.S., Bloomquist J.R., (2003). Differential up-regulation of striatal dopamine transporter and alpha-synuclein by the pyrethroid insecticide permethrin. *Toxicol. Appl. Pharmacol*. 192: 287–293.

Goldner W.S., Sandler D.P., Yu F., Hoppin J.A., Kamel F., Levan T.D. (2010). Pesticide use and thyroid disease among women in the Agricultural Health Study. *Am. J. Epidemiol.* 171: 455-64.

Hardell L., Eriksson M., Lenner P., Lundgren E. (1981). Malignant lymphoma and exposure to chemicals, especially organic solvents, chlorophenols, and phenoxy acids: a case-control study. *Br. J. Cancer.* 43: 169-176.

Harden L., Sandstrom A. (1979). Case-control study: soft tissue sarcomas and exposure to phenoxyacetic acids or chlorophenols. *Br. J. Cancer*. 39: 711.

Hart K., Pimentel D. (2002). Public health and costs of pesticides. In: Pimentel, D. (Eds.), *Encyclopedia of Pest Management*. Marcel Dekker, New York, pp 7-13.

Hoar S.K, Blair A., Holmes F.F., Boysen C.D., Robel R.J., Hoover R., Fraumeni J.F. (1986). Agricultural herbicide use and risk of lymphoma and soft tissue sarcoma. *JAMA*. 256: 1141-1147.

Hodge J.E. (1993). Pesticide Trends in the Professional and Consumer Markets. In: *Pesticides in Urban Environments.* Fate and Significance. Rackel K.D. and Leslie A.R. (eds) pp 10-17. American Chemical Society.

IARC (1991). Occupational exposures in insecticide application, and some pesticides. *IARC Monographs on the evaluation of the carcinogenic risk of chemicals to man*, Vol. 53. International Agency for Research on Cancer, Lyon.

IUPAC (International Union of Pure and Applied Chemistry). (2010). *History of Pesticide Use.* Available: http://agrochemicals.iupac.org/index.php?option=com_sobi2andsobi2 Task=sobi2Detailsandcatid=3andsobi2Id=31 [accessed 5 January 2012].

IUPAC (2011). International Union of Pure and Applied Chemistry. *Compendium of Chemical Terminology, Gold Book*. Version 2.3.

Jacobson J.L, Jacobson S.W. (2003). Prenatal exposure to polychlorinated biphenyls and attention at school age. *J. Pediatr*. 143: 780-788.

Johnson E.S. (1990). Association between soft tissue sarcomas, malignant lymphomas and phenoxy herbicides/chlorophenols: evidence from occupational cohort studies. *Fund. Appl. Toxicol.* 14: 219-234.

Kamel F., Hoppin J.A. (2004). Association of pesticide exposure with neurologic dysfunction and disease. *Environ. Health Perspect.* 112: 950-8.

Kelce W.R., Stone C.R., Laws S.C., Gray L.E., Kemppainen J.A., Wilson EM. (1995). Persistent DDT metabolite p,p'-DDE is a potent androgen receptor antagonist. *Nature.* 375: 581–585.

Kodavanti P.R. (2006). Neurotoxicity of persistent organic pollutants: possible mode(s) of action and further considerations. *Dose Response*. 3: 273-305.

Kolaczinski J.H., Curtis C.F. (2004). Chronic illness as a result of low-level exposure to synthetic pyrethroid insecticides: a review of the debate. *Food Chem. Toxicol.* 42: 697-706.

Lee D.H., Lee I.K., Jin S.H., Steffes M., Jacobs D.R. Jr. (2007). Association between serum concentrations of persistent organic pollutants and insulin resistance among nondiabetic adults: Results from the National Health and Nutrition Examination Survey 1999-2002. *Diabetes Care*. 30: 622-628.

Lee D.H., Lee I.K., Song K., Steffes M., Toscano W., Baker B.A., Jacobs D.R. Jr. (2006). A strong dose-response relation between serum concentrations of persistent organic pollutants and diabetes: Results from the National Health and Examination Survey 1999-2002. *Diabetes Care*. 29: 1638-1644.

Lee D.H., Lee I.K., Porta M., Steffes M., Jacobs D.R. (2007). Relationship between Serum Concentrations of Persistent Organic Pollutants and the Prevalence of Metabolic Syndrome among Nondiabetic Adults: Results from the National Health and Nutrition Examination Survey 1999–2002. *Diabetologia.* 50: 1841–1851.

Lee D.H., Jacobs D.R. Jr., Steffes M. (2008). Association of organochlorine pesticides with peripheral neuropathy in patients with diabetes or impaired fasting glucose. *Diabetes.* 57: 3108-11.

Montgomery M.P., Kamel F., Saldana T.M., Alavanja M.C., Sandler D.P. (2008). Incident diabetes and pesticide exposure among licensed pesticide applicators: Agricultural Health Study, 1993-2003. *Am. J. Epidemiol.* 167: 1235-1246.

Moreno G.D. (2003). *Toxicología Ambiental.* McGraw Hill, pp. 361.

Murphy S.D. (1986) Toxic effects of pesticides. In: *Casarett and Douls Toxicology: the basic science of poisons*, 3rd. edition, Klassen CD, Amdur MO and Doull J (eds), pp. 519-581. Macmillan, New York.

Nordberg M., Duffus, J.H., Templeton D., M. (2007). Explanatory dictionary of key terms in toxicology (IUPAC Recommendations 2007). *Pure Appl. Chem.* 79: 1583-1633.

Padungtod C., Savitz D.A., Overstreet J.W., Christiani D.C., Ryan L.M., Xu X. (2000). Occupational pesticide exposure and semen quality among Chinese workers. *J. Occup. Environ. Med.* 42:982–92.

Pearce N.E., Smith A.H., Howard J.K., Sheppard R.A., Giles H.J., Teague C.A. (1986). Non Hodgkin's lymphoma and exposure to phenoxy herbicides, chlorophenols, fencing work and meat work employment: a case-control study. *Br. J. Ind. Med.* 43: 75-83.

Peshin R. (2002). *Economic Benefits of Pest Management; Encyclopedia of Pest Management,* Pub. Marcel Dekker pp. 224-227.

Pimentel D. (2009). Pesticides and Pest Control. In: Rajinder Peshin and Ashok K. Dhawan (Editors). Integrated Pest Management: *Innovation-Development Process* Vol. 1. Springer, pp. 83-89.

Piña-Guzmán B., Solís-Heredia M.J., Quintanilla-Vega B. (2005). Diazinon alters sperm chromatin structure in mice by phosphorylating nuclear protamines. *Toxicol. Appl. Pharmacol.* 202: 189–98.

Piña-Guzmán B., Solís-Heredia M.J., Rojas-García A.E., Urióstegui-Acosta M., Quintanilla-Vega B. (2006). Genetic damage caused by methyl-parathion in mouse spermatozoa is related to oxidative stress. *Toxicol. Appl. Pharmacol.* 216: 216-24.

Porta M. (2006). Persistent organic pollutants and the burden of diabetes. *Lancet* 368: 558-559.

Purdue M.P., Engel L.S., Langseth H., Needham L.L., Andersen A., Barr D.B., Blair A., Rothman N., McGlynn K.A. (2009). Prediagnostic serum concentrations of organochlorine compounds and risk of testicular germ cell tumors. *Environ. Health Perspect.* 117: 1514-9.

Ray D.E., Richards P.G. (2001). The potential for toxic effects of chronic, low-dose exposure to organophosphates. *Toxicol. Terat*ol. 120: 343–351.

Richter E.D. (2002). Acute human pesticide poisonings. In: Pimentel, D. (Editor) *Encyclopedia of Pest Management*. Dekker, New York, pp. 3-6.

Ribas-Fitó N., Sala M., Cardo E., Mazón C., De Muga M.E., Verdú A., Marco E., Grimalt J.O., Sunyer J. (2002). Association of hexachlorobenzene and other organochlorine compounds with anthropometric measures at birth. *Pediatr. Res.* 52: 163-7.

Ribas-Fitó N., Torrent M., Carrizo D., Júlvez J., Grimalt J.O., Sunyer J. (2007). Exposure to hexachlorobenzene during pregnancy and children's social behavior at 4 years of age. *Environ. Health Perspect.* 115: 447-450.

Recio R., Ocampo-Gómez G., Morán-Martínez J., Borja-Aburto V., López-Cervantes M., Uribe M., Torres-Sánchez L., Cebrián M.E. (2005). Pesticide Exposure Alters Follicle-Stimulating Hormone Levels in Mexican Agricultural Workers. *Environ. Health Perspect.* 113: 1160–1163.

Rylander L., Rignell-Hydbom A., Hagmar L. (2005). A crosssectional study of the association between persistent organochlorine pollutants and diabetes. *Environ. Health.* 4: 28.

Sagiv S.K., Thurston S.W., Bellinger D.C., Tolbert P.E., Altshul L.M., Korrick S.A. (2010). Prenatal organochlorine exposure and behaviors associated with attention deficit hyperactivity disorder in school-aged children. *Am. J. Epidemiol.* 171: 593-601.

Salazar-Arredondo E., Solís-Heredia M.J, Rojas-García E., Hernández-Ochoa I., Quintanilla-Vega B. (2008). Sperm chromatin alteration and DNA damage by methyl-parathion, chlorpyrifos and diazinon and their oxon metabolites in human spermatozoa. *Repr. Toxicol.* 25: 455–460.

Sánchez-Peña L.C., Reyes B.E., López-Carrillo L., Recio R., Morán-Martínez J., Cebrián M.E., Quintanilla-Vega B. (2004). Organophosphorous pesticide exposure alters sperm chromatin structure in Mexican agricultural workers. *Toxicol. Appl. Pharmacol.* 196: 108–13.

Shafer T.J., Meyer D.A. (2004). Effects of pyrethroids on voltage-sensitive calcium channels: a critical evaluation of strengths, weaknesses, data needs, and relationship to assessment of cumulative neurotoxicity. *Toxicol. Appl. Pharmacol.* 196: 303–318.

Shafer T.J., Meyer D.A., Crofton K.M. (2005). Developmental neurotoxicity of pyrethroid insecticides: critical review and future research needs. *Environ. Health Perspect.* 113: 123-136.

Siddiqui M.K., Srivastava S., Srivastava S.P., Mehrotra P.K., Mathur N., Tandon I. (2003). Persistent chlorinated pesticides and intra-uterine foetal growth retardation: a possible association. *Int. Arch. Occup. Environ. Health.* 76: 75-80.

Skakkebaek N.E., Rajpert-De M.E., Main K.M. (2001). Testicular dysgenesis syndrome: an increasingly common developmental disorder with environmental aspects. *Hum. Reprod.* 16: 972–978.

Smith A.E., Bridges D.C. (1996). Movement of certain herbicides following application to simulated golf course greens and fairways. *Crop Sci.* 36: 1439-1445.

Smith J.G. (1988). Paraquat poisoning by skin absorption: a review. *Hum. Toxicol.* 7: 15-19.

Stephenson G.A., Solomon K.R. (1993). Pesticides and the Environment. In: The effects of pesticides in water resources. Aydinalp C., Porca M.M. 2004. *J. Central European Agriculture.* 5: 5-12.

Toppari J., Larsen J.C., Christiansen P., Giwercman A., Grandjean P., Guillette L.J. Jr, Jégou B., Jensen T.K., Jouannet P., Keiding N., Leffers H., McLachlan J.A., Meyer O., Müller J., Rajpert-De Meyts E., Scheike T., Sharpe R., Sumpter J., Skakkebaek N.E. (1996). Male reproductive health and environmental xenoestrogens. *Environ. Health Perspect* 104: 741–803.

Vallejo-Rosero M.C. (2006). Química de Plaguicidas. In: *Toxicología.* Córdoba-Palacio D. 5ta Ed. Bogotá. El Manual Moderno. 1044 pp.

Weisskopf M.G., Anderson H.A., Hanrahan L.P., Kanarek M.S., Falk C.M., Steenport D.M., Draheim L.A. (2005). Great Lakes Consortium. Maternal exposure to Great Lakes sport-caught fish and dichlorodiphenyl dichloroethylene, but not polychlorinated biphenyls, is associated with reduced birth weight. *Environ. Res.* 97: 149-62.

Weisskopf M.G., Knekt P., O'Reilly E.J., Lyytinen J., Reunanen A., Laden F., Altshul L., Ascherio A. (2010). Persistent organochlorine pesticides in serum and risk of Parkinson disease. *Neurology.* 74: 1055-61.

Wolff M.S., Anderson H.A., Britton J.A., Rothman N. (2007). Pharmacokinetic variability and modern epidemiology--the example of dichlorodiphenyltrichloroethane, body mass index, and birth cohort. *Cancer Epidemiol. Biomarkers Prev*. 16: 1925-30.

WHO (World Health Organization). (2004). *The WHO recommended classification of pesticides by hazard and guidelines to classification:* 2004. Available: http://www.who.int/ipcs/publications/pesticides_hazard_ rev_3.pdf [accessed 15 January 2012].

Zahm S.H., Blair A. (1992). Pesticides and Non-Hodgkin's Lymphoma. *Cancer Research Suppl*. 52: 5485-5488.

Zhang WenJun., FuBin Jiang., JianFeng Ou. (2011). Global pesticide consumption and pollution: with China as a focus. Proceedings of the International Academy of Ecology and Environmental. *Sciences*. 1: 125-144.

In: Pesticides: Characteristics, Uses and Health Implications
Editors: A. Abrego Gomez and E. Lugo de Ortega
ISBN: 978-1-62081-577-9

Chapter 8

Fate and Side-Effects of Pesticides in Urban Areas: A Case Study in Portugal

Michiel A. Daam**, Emília Silva, Ana Carina Santos Pereira, Luís M. Moreira da Silva, Ana Paula Ramos and Maria José Cerejeira

Instituto Superior de Agronomia/Technical University of Lisbon, CEER – Biosystems Engineering, Lisbon, Portugal

Abstract

The use of pesticides for pest control in non-agricultural or urban amenity areas has often been discussed to potentially lead to different fate and environmental as well as human side-effects than when used in agriculture. However, relatively little attention has been directed so far to the risk assessment of pesticides in urban as compared to agricultural areas, both in terms of research efforts and legislation. The present study was initiated as a first step to increase our knowledge related with pesticide use in urban areas. To this end, information on main pests, pesticides used to counteract these, and application techniques were gathered through surveys of pesticide operators in a Municipality in Portugal. A multicompartment model, leaching index and potential for particle-bound transport were applied by using the physicochemical properties of these pesticides to evaluate the environmental fate of the pesticides used in the present case study and obtain an indication of the compartments potentially at risk. These analyses will aid in selecting pesticides to be monitored in different compartments (drainage water, surface water, groundwater, soil) in future studies aiming at increasing our understanding of pesticide fate and their possible side-effects (applicatoroperator, bypassers, urban dwellers, environment) in urban settings. In addition, several constraints in knowledge of the operators related with pesticide application and storage were identified that may result in risk situations for both the environment and human health.

* Corresponding author: Michiel A. Daam, Instituto Superior de Agronomia / Technical University of Lisbon, Tapada da Ajuda, 1349-017 Lisbon, Portugal, Telephone: +351 21 365 3224 (direct), Fax: +351 21 365 3430, (mdaam@isa.utl.pt)

INTRODUCTION

Agriculture has long been regarded as the main source of pesticide inputs to surface waters. Over the last ten years, however, several studies have revealed that the usage and loss rates of pesticides in urban areas have been underestimated (e.g. Jongbloed et al., 2004; Gilliom et al., 2006; Kristoffersen et al., 2008; Hanke et al., 2010; Wittmer et al., 2011). For example, a monitoring study by Gilliom et al. (2006) reported that detection frequencies of pesticides in streams and groundwater of urban areas in the USA were very similar to detection frequencies in agricultural areas.

One of the reasons for the underestimation of potential risks related with non-agricultural applied pesticides is the lack of data on its use (Wittmer et al., 2011). For example, Kristoffersen et al. (2008) conducted a joint survey of seven European countries: Denmark, Finland, Germany, Latvia, the Netherlands, Sweden and United Kingdom. Statistics on pesticide use on urban amenity areas were only available in Denmark and the Netherlands (Kristoffersen et al. 2008).

Pesticide risk assessment and risk management have focused on agricultural usage, both in terms of scientific research efforts and political interest (Kristoffersen et al., 2008; Cohen, 2010). Our understanding of risks related with pesticide use in urban areas is therefore still relatively limited, both in terms of actual use and risks for the environment and humans. The present study was initiated to increase our understanding of pesticide use and related risks in urban areas. To this end, a survey among pesticide operators was conducted in a Municipality of Portugal to characterize the pesticides used for controlling pests, diseases and weeds occurring in public green spaces. A multicompartment model, leaching index and potential for particle-bound transport were applied by using the physicochemical properties of these pesticides to evaluate the environmental fate of the pesticides used in the present case study and obtain an indication of the compartments potentially at risk. In addition, information was gathered regarding the safety precautions taken before, during and after pesticide application, as well as knowledge of the operators related with these.

MATERIALS AND METHODS

Questionnaires

The study was conducted in a Municipality with over 150000 inhabitants in Portugal. The area contains a substantial number of urban green spaces (e.g., over 20 public gardens and playgrounds) and approximately 25% of the area covers parts of national reserve areas. Reliable records on pesticide applications were not available. Therefore, to create the pesticide inventory a questionnaire was submitted inquiring about pesticide application training and practices, and included asking-questions on procedures before (e.g., reading and understanding pesticide labels), during (e.g., use of protective clothing) and after (e.g., dealing with spray liquid remnants) pesticide application. Inquiries covered 24 out of a total number of 30 workers who mix and apply pesticides in this area, which corresponds to 80% of the pesticide operators.

A series of 13 chemicals was selected for a modelling approach. The list of commercial products applied and the organisms against which they were used are shown in Table 1.

Table 1. Main pesticides used in the Municipality under study and main phytosanitary problems against which these pesticides are used, as indicated by the pesticide operators. For *Bacillus thuringiensis*, no commercial product was indicated

Active ingredient	Commercial product	Pesticide type[1]	Formulation type[2]	Responded by % operators	Mainly used against
glyphosate-isopropylammonium	Cosmic; Nufosate	HB	SL; SL	42	weeds (streets, clay and gravel areas)
deltamethrin	Decis	IN	EC	42	broad-spectrum (caterpillars, aphids, scale insects)
2,4 D + MCPA	Bi-Hedonal 650	HB	SL	38	broad-leaved weeds (lawns)
glufosinate-ammonium	Basta S	HB	SL	25	weeds (streets, clay and gravel areas)
trichlorfon	Dipterex 80	IN	SP	25	caterpillars
paraquat	Paraquato Sapec 20	HB	SL	25	weeds (streets, clay and gravel areas)
chlorpyrifos	Dursban 5G; Pirifos 48	IN/AC	MG	21	broad-spectrum (caterpillars, aphids, scale insects)
diflufenican + glyphosate-isopropylammonium	Aliado	HB + HB	SC	17	weeds (streets, clay and gravel areas)
Propamocarb	Previcur N	FU	SL	17	downy mildew, powdery mildew, damping-off
2,4 D + mecoprop-P + dicamba	Trimec; Proscape	HB/PG + HB + HB	FG	17	broad-leaved weeds (lawns)
Bacillus thuringiensis	-	IN	-	17	pine processionary moth
cymoxanil + propineb	Milraz	FU + FU	WP	13	downy mildew, powdery mildew, damping-off
oxadiazon	Ronstar	HB	EC	13	broad-leaved weeds (lawns)

[1] HB = herbicide; IN = insecticide; AC = acaricide; FU = fungicide.

[2] SL = soluble concentrate; EC = emulsifiable concentrate; SP = soluble powder; MG = microgranules; SC = suspension concentrate; FG = fine granule; WP = wettable powder.

Level I Fugacity Model

Fugacity (Unit: Pa), the tendency of a chemical to escape from a phase (e.g., from water), has proven to be a convenient concept in environmental chemical equilibrium and partitioning calculations (Clark et al., 1988; Mackay, 1991). In the present study, the level I fugacity model version 3.00 (available via http://www.trentu.ca/academic/aminss/envmodel/models/L1300.html) described by Mackay (1991) was used to model the distribution of the pesticides used in the Municipality under study over the environmental compartments. This model assumes a simple, evaluative, closed environment with user-defined volumes and densities for the following homogeneous environmental media (or compartments): air, water (without distinction between surface water and groundwater), soil, sediment, suspended sediment, fish, and aerosols. A Level I calculation gives a general impression of the likely media into which a chemical will tend to partition and an indication of relative concentrations in each medium. Physical-chemical properties of the chemicals of concern are used to quantify a chemical's behavior in this evaluative environment. Three types of chemicals are treated in level I simulations of this model: chemicals that partition into all media (Type 1), involatile chemicals (Type 2), and chemicals with zero, or near-zero, solubility (Type 3). The model was run at default settings with input values for the physical-chemical parameters as provided in the Footprint pesticide properties database (accessible via: http://sitem.herts.ac.uk/aeru/footprint/en/index.htm; Table 2). If more than one value was available, the average value was used. Since the value for log Kow that can be entered in the model should be between -3 and 12, -3 was used in cases of a log Kow < -3. For deltamethrin and paraquat, the vapour pressure values as indicated in the Footprint database differed substantially from those reported in their respective EU review reports (EC, 2002, 2003b) among other data sources as indicated by Batista (2003). Subsequently, input and output data of the model calculations as reported in Batista (2003) are presented in Table 2 and 3, respectively.

Groundwater Ubiquity Score (GUS)

The potential of the compounds under study to leach to groundwater ecosystems was evaluated by determining the Groundwater Ubiquity Score (GUS). The GUS is calculated based on the two main properties of pesticides that determine leachability, that is, the Soil Organic Carbon partition Coefficient (Koc) and the soil half-life (DT50), according to the following formula (Gustafson, 1989):

$$GUS = \log 10 (DT50) \times (4 - \log 10 (Koc))$$

The value set at 4 in the formula is arbitrary selected to permit separation of substances into three categories: nonleachers (GUS < 1.8), transition (1.8 < GUS < 2.8), and leachers (GUS > 2.8) (Gustafson, 1989).

Potential for Particle-Bound Transport

The potential for particle-bound transport was based on the DT50 in soil, Koc and water solubility of the pesticides, using a classification system described by Spikkerud (2002):

- High potential: DT50 > 40 days and Koc >1000
  DT50 > 40 days, Koc > 500 and solubility < 0,5 mg/l
- Low potential: DT50 < 1 day
  DT50 < 2 days and Koc < 500
  DT50 < 4 days, Koc < 900 and solubility > 0,5 mg/l
  DT50 < 40 days, Koc < 500 and solubility > 0,5 mg/l
  DT50 < 40 days, Koc < 900 and solubility > 2 mg/l
- Medium: All other

Table 2. Physico-chemical input parameters used for the level I fugacity model of Mackay (1991) and calculation of the Groundwater Ubiquity Score (GUS). Data were obtained from the Footprint pesticide properties database (available via: http://www.eu-footprint.org/ppdb.html; accessed in January 2012) unless otherwise indicated

Active ingredient	Molecular mass (g/mol)	Melting point (°C)	Solubility in water at 20°C (mg/L)	Vapour pressure at 25°C (mPa)	Henry's law constant at 25°C (Pa m^3 mol^{-1})	Log K_{ow} at pH 7, 20°C	K_{oc} (mL/g)	pK_a (or pK_b) at 25°C	DT_{50} soil (days)[1]
Herbicides/plant growth regulators									
2,4 D	221	139	2.3E+04	1.9E-02	1.3E-05	-0.83	1.7E+02	2.9	10
dicamba	221	115	2.5E+05	1.7	1.0E-04	-1.9	12	1.9	8
diflufenican	394	160	5.0E-02	4.3E-03	1.2E-02	4.2	2.0E+03	(3)	1.8E+02
glyphosate-isopropylammonium[8]	228	206	1.2E+04	2.1E-03	4.6E-10	-5.4	6.9E+03	4.0	59
glufosinate-ammonium	198	217	5.0E+05	3.1E-02	4.5E-09	-4.0	6.0E+02	9.2	7.4
MCPA	201	116	2.9E+04	0.4	5.5E-05	-0.81	74	3.7	15
mecoprop-P	215	95	8.6E+02	0.23	5.7E-05	0.020	31	3.9	8
oxadiazon	345	89	0.57	6.7E-01	3.8E-02	5.3	2.2E+03	(3)	5.0E+02
paraquat*	257	340[4]	6.2E+05	0	4.0E-12	-4.5	1.0E+06	(3)	3.0E+03
Insecticides / acaricides									
Bacillus thuringiensis	35792	(7)	10	(7)	(7)	(7)	5.0E+03	(7)	2.7
chlorpyrifos	351	42	1.1	1.4	0.48	4.7	8.2E+03	(3)	50
deltamethrin*	505	100	2.0E-03	1.0E-05	3.1E-02	4.6	4.6E+05	(3)	< 23
trichlorfon	257	80	1.2E+05	0.21	4.5E-07	0.43	10	(6)	18
Fungicides / bactericides									
cymoxanil	198	161	7.8E+02	0.15	3.8E-05	0.67	44	9.3	0.7
propineb	290	150[2]	10	0.16	8.0E-08	-0.26	0.23[2]	(2)	3
propamocarb	188	64.2[5]	9.0E+05	7.3E+02	1.5E-04	0.84	7.1E+02[5]	9.5	39[5]

* Presented values are those used by Batista (2003; for rationale, see text).

[1] The "typical" DT_{50} value (i.e. mean of all field and laboratory studies) provided in the FOOTPRINT database was used.

[2] Propineb decomposes before melting and Koc and pKa are not applicable due to insolubility and instability in water (EC, 2003a). To enable running the McKay model, the degradation point was used to set the melting point at 150°C and Koc was calculated using the equation: Koc = 0.41 Kow (after Karickhoff, 1981).

[3] No dissociation as indicated in EC (2006a) for diflufenican and in the Footprint database for the other compounds.

[4] The technical concentrate shall consist essentially of an aqueous solution of paraquat dichloride and paraquat dichloride decomposes at approximately 340 °C (EC, 2003b). To enable running the McKay model, this value was used as the value for the melting point.

[5] Propamocarb is normally used as the hydrochloride salt in formulated products (EFSA, 2006). In the absence of melting point, Koc, and DT_{50} data for propamocarb, data for propamocarb hydrochloride available in the Footprint database were used here to enable running the McKay model and calculating the GUS.

[6] Trichlorfon exhibits no basic properties in aqueous solution. Acidic properties can not be determined, since deprotonation followed by formation of DDVP occurs in the presence of bases. It is not possible to specify a pKa value for trichlorfon in water (EC, 2005b).

[7] *Bacillus thuringiensis* is a bacterial insecticide so several physico-chemical parameters do not apply. Subsequently, the MacKay model was not run for *B. thuringiensis*.

[8] Data missing in the FOOTPRINT (no data in EU database) were taken from Tomlin (2006).

Results And Discussion

Pesticides Used and Phytosanitary Problems

A total number of 15 commercial products (not counting with *Bacillus thuringiensis*, for which no commercial product was indicated) were indicated by the operators to be applied in the study area (Table 1). The type of pesticides mostly mentioned to be used was herbicides (57%), followed by insecticides (34%) and fungicides (9%). This is in agreement with the type of plant enemies mostly indicated: weeds (45%), insects (37%) and fungal diseases (18%). Main weed problems were subdivided in two subgroups: i) weeds on streets, and clay and gravel areas, with products containing glyphosate-isopropylammonium, glufosinate-ammonium and paraquat indicated mostly to counteract these, and ii) broad-leaved weeds on lawns, with Bi-Hedonal 650 as the product mostly mentioned to be used against these (Table 1). A variety of pest problems on herbaceous plants, shrubs and trees were indicated: caterpillars, aphids and scale insects; the pine processionary moth (*Thaumetopoea pityocampa*) was mentioned separately from the other groups of insects. The most frequently mentioned insecticide was Decis (active ingredient deltamethrin) and the bioinsecticide *Bacillus thuringiensis* was the only pesticide used in the study area against the pine processionary moth (Table 1), although other substances are approved for the control of this pest (DGADR, 2012). Two fungicides were pointed out, cymoxanil+propineb and propamocarb (hydrochloride). Although these fungicides are certified for use under integrated pest management programs, the first one is not authorized for urban uses and the propamocarb can only be applied in soil applications to control damping-off and root rots caused by *Pythium* and *Phytophthora* fungi (DGADR, 2012)

Regarding application frequency, most spots are applied twice a year (79%), whereas a single (8%) or three applications (13%) per year are less common. Unfortunately, applicators were not asked how often and in how many places they apply the different indicated pesticides individually so no quantitative evaluation of the individual pesticides could be made (e.g., a pesticide could hypothetically be applied once a year by many applicators in a few places or by a few applicators in many places with a greater frequency). From previous studies, however, it appears that herbicides indeed constitute the main part of the pesticides used in urban areas (e.g., Kristoffersen et al., 2008; Hanke et al., 2010).

Despite that trichlorfon and paraquat were taken from the European (by 2007/356/EC and Judgment of the Court of First Instance 11/07/2007, respectively) and national market (DGADR, 2008) in 2007, their use was indicated by one in four operators (Table 1). This use could reflect a historical rather than an actual use since surveys were conducted at the end of 2010/beginning 2011, whereas the national stock depletions at the user were indicated for 31/12/2008 (paraquat) and 21/05/2009 (trichlorfon) (DGADR, 2012).

Relevant is also the fact that none of the operators that indicated to use trichlorfon and paraquat appeared to be aware of both its use cancellation and that some of the active ingredients used otherwise were not approved for the intended purpose, even after informing them about this. Likewise, the lack of knowledge of the identity of the harmful organisms to be controlled, the use of inadequate or non-authorized pesticides, and the application of unselective treatments, are procedures that can negatively influence the populations of natural enemies and possibly lengthen the pest outbreaks, making it more severe and problematic to control.

Pesticide Application Practices and Potential Human Health Risks

All commercial products indicated as used in the study area except Dursban 5G (microgranules) and Proscape Extra (fine granule that also contains fertilizer) are products with formulations for mixing with water that is applied by pulverization (Table 1). In line with this, application by motorized (69%) and knapsack (27%) sprayers were indicated to be used in the vast majority of applications. The motorized sprayers (air blast sprayers) are transported in vehicles and pesticide applications are usually made without taking the sprayer from the vehicle (pers. comm. with operators). Although this is evidently the most efficient in terms of time load, it also implies that little care is taken to adequately calibrate the sprayers.

Various questions were made regarding the safety precautions taken before, during and after pesticide application. An overview of selected results from these questions is scrutinized below.

- *Consideration of information on pesticide label*: Although 83% of the respondents answered that they read the pesticide label and understand the toxic symbols, 21% responded that they do not follow the instructions on the pesticide labels and 38% indicated that they do not consider the information on the labels important/relevant.
- *Pesticide spraying liquid preparation (Figure 1)*: Most of the operators said they prepare the pesticide spraying liquid onsite (92%). Although 71% indicated that they calculate the pesticide dose, with the remaining 29% estimating the amount to be "applied by eye", only 58% answered that they use graduated containers to this end.

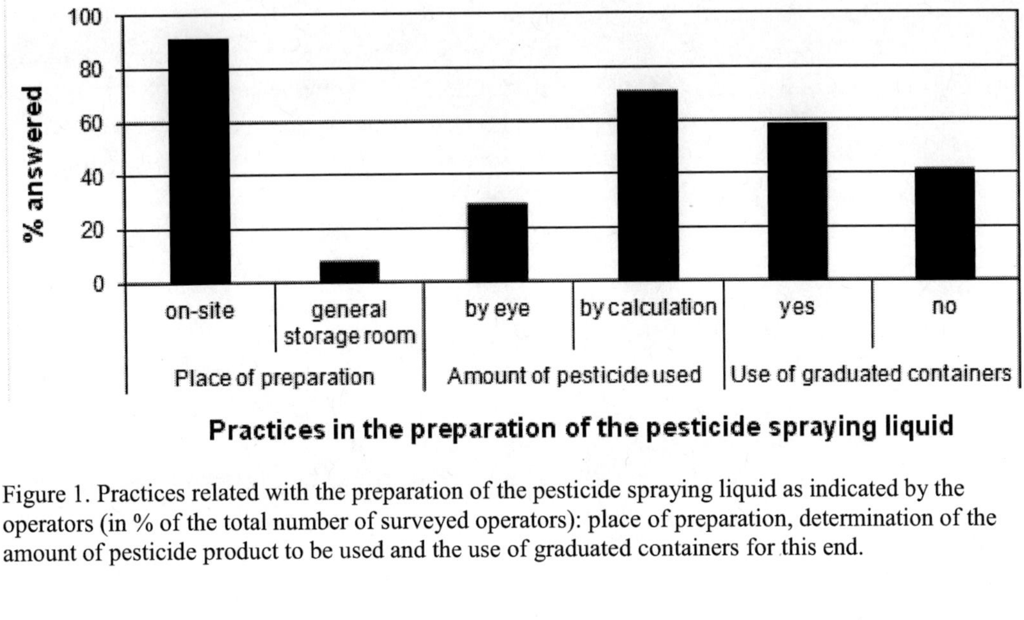

Figure 1. Practices related with the preparation of the pesticide spraying liquid as indicated by the operators (in % of the total number of surveyed operators): place of preparation, determination of the amount of pesticide product to be used and the use of graduated containers for this end.

- *Use of protection equipment (Figure 2):* All operators indicated that they use gloves during preparation of the pesticide spraying liquid and application. Despite that all indicated to use a mask during pesticide application, this was only 75% during preparation of the pesticide spraying liquid. Use of safety glasses (around half), impermeable boots (one in three) and face shield (one to four out of 24) appear to be very low.

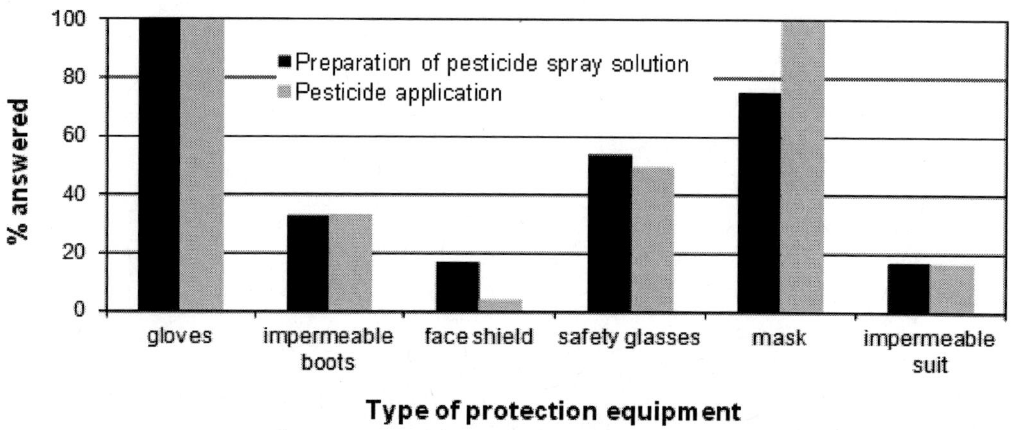

Figure 2. Use of protection equipment (gloves, impermeable boots, face shields, safety glasses, mask, and impermeable suit) during preparation of the pesticide spraying liquid and application as indicated by the operators (in % of the total number of surveyed operators).

- *Proximity of application to humans, animals and water*: A great number of operators indicated that they apply pesticide close to people (83%), animals (67%), schools (48%) and watercourses (28%).
- *Warning signs after application*: All operators indicated that they do not put warning signs before, during or after application.
- *Pesticide spray liquid remnants*: After applying the pesticide, spray liquid remnants were indicated to be stored (25%), pulverized over the treated area (42%), pulverized over a non-treated area (21%), released in watercourses, domestic drainage system or storm sewer (13%). Evidently, the amount of spray liquid remnants should be minimized by properly calculating the dose and by adequately calibrating the sprayer. Based on the discussed above, it may at least be questioned whether this is done accurately by the surveyed operators. Also, only 21% pointed out that they treat the spray liquid remnants according to the appropriate procedure, i.e. pulverization over a non-treated area containing vegetation (after diluting the spray liquid remnant; TOPPs, 2008). It should be noted, however, that "pulverized over the treated area" and "stored" could be accurate in case of under-dosed zones and re-utilization of stored spray liquid remnants within 24 hours, respectively (TOPPs, 2008).
- *Rinsing of spraying equipment*: The spraying tank was indicated to be washed once (25%), twice (42%) or three (33%) times, and the rinsing water released in a non-treated area (29%), storm sewers (25%), particular area dedicated for this (17%), treated area (13%), domestic drainage system (13%), or ditches (4%). The spraying

tank should be rinsed three times and released over a non-treated area containing vegetation (TOPPs, 2008).

- *Fate of empty pesticide containers*: When asked how they dispose of empty pesticide containers, the operators responded bury (4%), burn (13%), utilize for other purposes (17%; mentioned were e.g. for fuel and garbage), and urban waste containers (50%). The correct way, i.e. deliver in recycle points for empty pesticide containers (in Portugal: Valorfito, http://www.valorfito.com/) was only answered by 17 % (4 out of 24).
- *Storage*: Only 21% of the operators indicated that a storage room was exclusively used for storing of pesticides. In line with this, only one of the four storage rooms (i.e., 25%) used in the Municipality had a room exclusively designated for storage of pesticides. In another situation, pesticide products were stored without pesticide labels in an area accessible to people and animals. The absence of pesticide labels was indicated to be due to the fact that the content of the original package are divided over various other containers, which are used for transportation to the application site.

The risk situations described above may be partly related with the low level of training in pesticide application practices since 58% of the operators indicated that they never had such training. The remaining 42% indicated that they participated in a basic training of 35 hours, covering identification of phytosanitary problems, application procedures and some precautions when using pesticides, though environmental issues were indicated to be hardly covered. The reason that was given for not having followed the training was that at the time this was provided, their duties did not include the application of pesticides. Training of operators is essential to mitigate risks related with the use of pesticides (e.g. TOPPs, 2008). On legislative level, improving training of operators has also been acknowledged in the EU to be a condition to reduce potential risks to human health and the environment as to achieve a sustainable use of pesticides (EU Directive 2009/128/EC; EC, 2009).

The potential human risks related with the incorrect/inappropriate practices described above may be illustrated with the EU and WHO toxicity classification, and the EU risk and safety phrases of the active ingredients used in the commercial products that are utilized in the study area (Table 4). For example, glyphosate-isopropylammonium is classified as slightly hazardous by WHO and irritant/dangerous for the environment in the EU. The risk phrase R41 indicates "risk of serious damage to eyes" and subsequently, special attention should be paid to "wear eye/face protection" (S39; EC, 2001). Despite this, the actual use of safety glasses (around half) and face shield (one to four out of 24) was indicated to be very low. The acceptable operator exposure level (AOEL) is the maximum amount of active substance to which the operator may be exposed via all relevant exposure routes (inhalation, dermal, and, to a lesser extent, oral) without any adverse health effects (EC, 2006b), and can hence be used as an indicator for its toxic potency to operators. Consequently, from the information contained in Table 4, it can be concluded that especially diflufenican, glufosinate-ammonium, paraquat, deltamethrin, and propineb may exert toxic effects to the operators at very low (internal) body doses. It should be noted that AOELs are also set to ensure no harmful effects on the health bystanders and residents, besides the operators and workers (EC, 2006b). As discussed above, applications were indicated to be made close to people (83%), animals (67%), and schools (48%) and no signs are put after application, which may thus indicate potential risks for human health besides that of the operators. In addition, the risk phrases "keep locked up" (S1) and "keep out of the reach of children" (S2) are not met in the storage room that was discussed above to be accessible to people and animals.

Potential Environmental Risks

The predicted environmental behavior of the pesticides indicated as used in the Municipality under study are presented in Table 3. As can be deducted from this Table, herbicides and insecticides tended to have in general the greatest affinity for the water and soil compartments, respectively. After reviewing pesticide monitoring studies in Portuguese groundwater and surface water, Cerejeira et al. (2003), Silva et al. (2006) and Daam et al. (2011) indeed reported that both the number and concentrations of herbicides encountered were greater than those of insecticides. They attributed this to the greater water solubility of the herbicides as compared to insecticides that were included in the analyses (Daam et al., 2011). In line with this, the herbicides diflufenican and oxadiazon have low water solubility values as compared to the other herbicides and a low affinity to the water compartment (Tables 2 and 3). The only insecticide for which a high PEDwater was calculated, i.e. trichlorfon, also has high water solubility. For the herbicide paraquat, however, a PEDwater of zero was calculated for the water compartment despite a high value for water solubility (Tables 2 and 3). This may be explained by the high value for Koc, indicating that paraquat has a great affinity to the organic material in soil and will thus not be easily available to dissolve and subsequently transported in runoff water. On the other hand, this also indicates a high potential for particle-bound transport, i.e. transported in runoff water bound to soil particles (Table 3). Since level I simulations of the fugacity model do not consider advection processes (Mackay, 1991), this means that amounts of this compound being transported with water in real-world conditions may be underestimated with these simulations. The same holds true for the other pesticides for which low PEDwater and high potential for particle-bound transport are indicated in Table 3. Hence, though the present analyses provides indications for which pesticides should be considered in monitoring studies of soil and water, this should be confirmed and eventually adjusted based on results from field studies.

The need for studies into monitoring of pesticides in urban areas has previously been stressed, the more as the use of pesticides in non-agricultural areas has been reported to potentially lead to different environmental issues than when they are used in agriculture (Spliid et al., 2004; Kristoffersen et al., 2008). For example, hard surfaces in urban areas are often constructed for rapid penetration of water or to encourage surface runoff to avoid flooding, which may result in contamination of nearby ditches, drains, sewage systems or groundwater (Kristoffersen et al., 2008). Consequently, there is minimal opportunity for pesticide sorption and/or degradation, while the potential for pesticide removal to surface water bodies is high (Ramwell et al., 2002). Additionally, it has been demonstrated for glyphosate that besides run-off, leaching of pesticides from hard surfaces into groundwater cannot be ignored, depending on weather and joint conditions (Kristoffersen et al., 2008). The discussed rapid penetration and runoff of water in hard surfaces also imply high pesticide loss rates. Wittmer et al. (2011) recently conducted a comprehensive field study in a catchment of mixed land use in Switzerland to compare pesticide use and loss rates from urban and agricultural origins. Despite the substantially lower amounts used, the measured loads of urban pesticides were in the same range as the most widely-used agricultural pesticides. The lower usage was explained by the authors to be compensated by urban loss rates that were up to ten times higher than for agricultural ones (0.6 to 15% for urban, 0.4 to 0.9% for agricultural compounds; Wittmer et al., 2011). In addition, the removal of pesticides by treatment plants has been discussed to be costly and to have low efficiencies (e.g., Puijker et al., 2004; Van de Steene et al., 2010)

Table 3. Predicted environmental distributions (PEDs; in %) of the pesticides over the different environmental compartments as calculated by the level I fugacity model of Mackay (1991), and their potential to leach to groundwater as expressed by the Groundwater Ubiquity Score (GUS). Since PEDs in water and soil combined were close to 100% for all compounds, only these PEDs are shown and PEDs ≥ 1% for other compartments indicated. GUS classifications were made according to Gustafson (1989): nonleachers (GUS < 1.8); transition (1.8 < GUS < 2.8); and leachers (GUS > 2.8). For *Bacillus thuringiensis*, the fugacity model was not run (NR; for rationale, see Table 2)

Pesticide	PED water	PED soil	GUS	GUS classification	Potential for particle bound transport
Herbicides/plant growth regulators					
2,4 D	100	0	1.8	transition	low
dicamba	99	0	2.6	leacher	low
diflufenican[1]	7	91	1.6	non-leacher	high
glyphosate-isopropylammonium	100	0	-0.3	non-leacher	high
glufosinate-ammonium	100	0	1.1	non-leacher	low
MCPA	100	0	2.5	leacher	low
mecoprop-P	100	0	2.3	leacher	low
oxadiazon	1	97	1.8	transition	high
paraquat[*,1]	0	98	-7.0	non-leacher	high
Insecticides / acaricides					
Bacillus thuringiensis	NR	NR	0.1	non-leacher	medium
chlorpyrifos[1]	2	96	0.1	non-leacher	high
deltamethrin[*,2]	3	94	-2.3	non-leacher	medium
trichlorfon	100	0	3.8	leacher	low
Fungicides / bactericides					
cymoxanil	100	0	-0.4	non-leacher	low
propineb	100	0	2.2	leacher	low
Propamocarb	99	1	1.8	transition	medium

[*] Presented values are those calculated by Batista (2003; for rationale, see text).

[1] PED sediment = 2%.

[2] PED sediment = 2%; PED air = 1%.

Table 4. Indication of the human and environmental toxicity classification, safety phrases and AOEL (Acceptable Operator Exposure Level - Systemic) of the identified active ingredients as set in the EU (Source: Footprint database; available via: http://www.eu-footprint.org/ppdb.html; accessed in January 2012). The WHO (World Health Organization) classification is also provided for comparison. For a list of the EU risk and safety phrases, please refer to Comission Directive 2001/59/EC (EC, 2001; available via http://eur-lex.europa.eu/JOHtml.do?uri=OJ:L:2001:225:SOM:EN:HTML)

Active ingredient (WHO classification)[1]	Human risk classification[2]	Safety phrases	AOEL (mg kg^{-1}bw day^{-1})	Environmental classification[2]
Herbicides/plant growth regulators				
2,4 D (II)	Xn - R22; Xi - R37, R41, R43	S2, S24/25, S26, S46, S61, S36/37/39	0.15	N - R52, R53
dicamba (II)	Xn - R22; Xi - R41	S2, S22, S61	0.3	N - R52, R53
diflufenican (III)	-[3]	S60, S61[3]	0.0033	N - R52/53[3]
glyphosate-isopropylammonium (III)	Xi - R41	S2, S26, S39, S61	-	N - R51, R53
glufosinate-ammonium (II)	R60, R63[4]; Xn - R20/21/22, R48/20/22	S45, S53	0.0021	-
MCPA (II)	Xn - R22; Xi - R38, R41	S2, S26, S37, S39, S60, S61	0.04	N - R50, R53
mecoprop-P (II)	Xn - R22, Xi - R41	S2, S13, S26, S37/39, S60, S61	0.04	N - R51, R53
oxadiazon (U)	R62[5]	S60, S61	0.05	N - R50, R53
paraquat (II)	T+ - R26; T - R24/25, R48/25; Xi - R36/37/38	S1/2, S22, S28, S36/37/39, S45, S60, S61	0.0004	N - R50, R53
Insecticides / acaricides				
Bacillus thuringiensis (III)	Xi - R43[6]	S2, S13, S20/21, S24/25, S26[6]	-[6]	-[6]
chlorpyrifos (II)	T - R25	S1/2, S45, S60, S61	0.01	N - R50, R53
deltamethrin (II)	T - R23/25	S1/2, S24, S28, S36/37/39, S38, S45, S60, S61	0.0075	N - R50, R53
trichlorfon (II)	Xn - R22; Xi - R43	S2, S24, S37, S60, S61	0.09	N - R50, R53
Fungicides / bactericides				
cymoxanil (II)	Xn - R22; Xi - R43	S2, S36/37, S60, S61	0.01	N - R50, R53
propineb (U)	Xn - R20, R48/22; Xi - R43	-[7]	0.003	N - R50
propamocarb (U)	Xn - R22	S2, S13, S24, S37, S46, S61[6]	0.45[6]	N - R52[8]

[1] WHO classification: Class Ia: extremely hazardous; Class Ib: highly hazardous; Class II: moderately hazardous; Class III: slightly hazardous; O: Obsolete substance; NL: Not listed; U: Unlikely to present an acute hazard (Source: Footprint database).

[2] EU toxic symbols: T = toxic; T+ = very toxic; Xn = harmful; Xi = irritant; N = dangerous for the environment.

[3] For diflufenican, no information was provided on the Footprint database regarding the EU data; reported data were obtained from the draft assessment report of this active ingredient (EC, 2006a).

[4] Reproduction risk category 2 and 3.

[5] Reproduction risk category 3.

[6] For *Bacillus thuringiensis*, no information was provided on the Footprint database regarding the EU data; reported data were obtained from the draft assessment report of *Bacillus thuringiensis* subsp. Aizawai - strain ABTS-1857/Xen Tari WG (EC, 2008).

[7] No information on safety phrases, neither in the Footprint database nor in EU review report (EC, 2003a).

[8] For propamocarb, no information was provided on the Footprint database regarding the EU data; reported data were obtained from the draft assessment report of this active ingredient (EC, 2005a).

Pesticides used in non-agricultural areas were primarily developed for use on cropland because the non-agricultural market is small; still they are covered by the same pesticide regulation and have to meet similar criteria (Kristoffersen et al., 2008). From the above, it may be at least questionable whether the scenarios used by FOCUS (FOCUS, 2001) to estimate predicted environmental concentrations in water bodies in agroecosystems are valid for urban conditions. Monitoring pesticide concentrations in urban areas and subsequently calibrating and validating model scenarios seems a meaningful direction to follow. These scenarios should consider the various application location types and the variety within these. For example, the applicators pointed out that they apply pesticides in public gardens (63%), streets (50%) and lawns (46%). Furthermore, indicated substrates onto which pesticides were applied differed between these location types: unpaved, paved and tar were mentioned by respectively 17%, 88%, and 8% for streets, and 54%, 63%, 4% for gardens. Evidently, differences in environmental/physicochemical conditions of these substrates as well as the weather conditions should also be considered. Regarding weather conditions, 92% of the applicators indicated that they apply pesticides during calm weather, whereas none and only one (out of 24) responded that they apply during rain or windy weather, respectively. It thus appears that weather conditions are relatively well considered to avoid runoff (by rain) and spray drift (by high wind speeds). When questioned about the time of the day pesticides are applied, the respondents answered morning (100%), afternoon (29%) and evening/night (67%). Although applications in the morning and afternoon are understandable for a logistic point of view, the chance of bystanders is evidently lowest during the night. Furthermore, morning applications are less favorable in case of photodegradable pesticides since they will be immediately and prolonged subjected to sunlight. In the survey, 33% of the applicators indeed indicated to apply pesticides in sunny conditions. In case of flower gardens, afternoon applications may also be less favorable in case pesticides with potential side-effects on bees are applied because bees forage in the afternoon (Marer, 1999).

All active ingredients except *B. thuringiensis* are classified as "dangerous for the environment" (N), and subsequently include the risk phrases R50 (Very toxic to aquatic organisms), R51 (Toxic to aquatic organisms), R52 (Harmful to aquatic organisms) or R53 (may cause long-term adverse effects in the aquatic environment), and the safety phrase "avoid release to the environment" (EC, 2001). Despite this, a great number of operators indicated that they apply pesticides close to watercourses (28%), release pesticide spray liquid remnants in watercourses, domestic drainage system or storm sewer (13%), and release the rinsing water of spraying equipment in storm sewers (25%), domestic drainage system (13%), or ditches (4%).

CONCLUSION

Based on these study findings, it appears there is an urgent need to increase the level of training, e.g. like developed by TOPPs, to pesticide operators in urban areas. Furthermore, monitoring studies in urban areas should be conducted to evaluate whether humans, animals and environment are at risk, even if appropriate practices are applied. Besides to validate whether the compartments indicated in the present study have indeed the greatest potential for pesticide contamination, this also includes the development of fate scenarios for urban areas.

Eventually, this could imply that fate of pesticides applied in urban areas should be better included in the risk assessment of pesticides, which is currently focussed on agricultural scenarios.

According to forecasts, up to two-thirds of the world's population will live and work in urban centres in only 20 to 50 years from now. The demand for public green spaces has risen as they form essential parts of cities and substantially determine the city-dwellers' quality of life. Gradually we are witnessing both a rising public sense for environmental questions and a growing commitment of city managers to keep plants in healthy conditions. However, the use of pesticides for pest control in urban environments will be more limited in the coming years due to legislative restrictions. In a near future, more emphasis should be placed on the formulation of joint approaches to knowing and solving phytosanitary problems affecting green urban spaces, on the use of more environmental friendly pesticides as part of integrated pest strategies in plant health programs and on the development of integrated environmental programs for the assessment of pesticides' fate in urban environments.

Acknowledgment

The authors are indebted to the Municipal Hall and pesticide operators of the Portuguese Municipality under study for their kind co-operation during this study.

References

Batista S (2003). *Exposição de águas subterrâneas de ecossistemas agrícolas a pesticidas* ("groundwater exposure to pesticides in agroecosystems"). Doctoral thesis Technical University of Lisbon, Portugal (in Portuguese).

Cerejeira MJ, Viana P, Batista S, Pereira T, Silva E, Valério MJ, Silva A, Ferreira M, Silva-Fernandes AM (2003). Pesticides in Portuguese surface and ground waters. *Water Res.* 37: 1055-1063.

Clark T, Clark K, Paterson S, MacKay D, Norstrom RJ (1988). Wildlife monitoring, modeling and fugacity. *Environ. Sci. Technol.* 22: 120–127.

Cohen SZ (2010). Urban pesticide risk assessment and risk management: get involved. *Environ. Toxicol. Chem.* 29: 1201-1202.

Daam MA, Silva E, Leitão S, Cerejeira MJ (2011). Environmental risk assessment of pesticides in Mediterranean Portugal: Status and research needs. In: Daniels JA (ed.). *Advances in Environmental Research. Volume 9*. Nova Science Publishers, New York. pp. 213-238.

DGADR (2008). Retirada de substâncias activas no mercado e sua consequência na protecção das culturas. Direcção-Geral de Agricultura e Desenvolvimento Rural (DGADR). Available via: *http://infoagro.cothn.pt/files/_Retirada_48b7fd92b8243.pdf.*

DGADR (2012). *Produtos fitofarmacêuticos. Direcção-Geral de Agricultura e Desenvolvimento Rural* (DGADR; in Portuguese). Available at *http://www.dgadr.pt/* (accessed in February 2012).

EC (2001). Commission Directive 2001/59/EC of 6 August 2001 adapting to technical progress for the 28th time Council Directive 67/548/EEC on the approximation of the laws, regulations and administrative provisions relating to the classification, packaging and labelling of dangerous substances. *Off. J. Eur. Comm.* L225: 1-333.

EC (2002). Review Report for the active substance deltamethrin. SANCO/6504/VI/99-final. *European Commission (EC),* Brussels, Belgium.

EC (2003a). Review Report for the active substance propineb. *SANCO/7574/VI/97-final. European Commission (EC),* Brussels, Belgium.

EC (2003b). Review Report for the active substance paraquat. *SANCO/10382/2002 -final. European Commission (EC),* Brussels, Belgium.

EC (2005a). Draft Assessment Report - public version. Propamocarb. Volume 3, *Annex B, B.1 - B.5 & B.6 part 4. European Commission (EC),* Brussels, Belgium.

EC (2005b). Draft Assessment Report - public version. Trichlorfon. Volume 1. *European Commission (EC),* Brussels, Belgium.

EC (2006a). Draft Assessment Report - public version. Diflufenican. Volume 3, Annex B, B.1 - B.5. *European Commission (EC),* Brussels, Belgium.

EC (2006b). Draft Guidance for the setting and application of acceptable operator exposure levels (AOELs). SANCO 7531 - rev.10. *European Commission (EC),* Brussels, Belgium.

EC (2008). Draft Assessment Report - public version. *Bacillus thuringiensis* subsp. Aizawai - strain ABTS-1857/Xen Tari WG. Volume 1. *European Commission (EC),* Brussels, Belgium.

EC (2009). Directive 2009/128/EC of the European Parliament and of the Council of 21 October 2009 establishing a framework for Community action to achieve the sustainable use of pesticides. *Off. J. Eur. Comm.* L309: 71-86.

EFSA (2006). Conclusion regarding the peer review of the pesticide risk assessment of the active substance propamocarb. EFSA (European Food Safety Authority) Scientific Report 78: 1-80.

FOCUS (2001) FOCUS surface water scenarios in the EU evaluation process under 91/414/EEC. Report of the FOCUS (Forum for the Co-ordination of Pesticide Fate Models and Their Use) Working Group on Surface Water Scenarios. EC Document Reference SANCO/4802/2001-rev.2.

Gilliom RJ, Barbash JE, Crawford CG, Hamilton PA, Martin JD, Nakagaki N, Nowell LH, Scott JC, Stackelberg PE, Thelin GP, Wolock DM (2006). The quality of our nation's waters - pesticides in the nation's streams and groundwater, 1992– 2001. Reston (VA): US Geological Survey. *Circular* 1291, Sacramento, USA.

Gustafson DI (1989). Ground water ubiquity score: A simple method for assessing pesticide leachability. *Environ. Toxicol. Chem.* 8: 339-357.

Hanke I, Wittmer I, Bischofberger S, Stamm C, Singer H (2010). Relevance of urban glyphosate use for surface water quality. *Chemosphere* 81: 422-429.

Jongbloed RH, Hulskotte JHJ, Kempenaar C (2004). Contribution of agricultural and non-cultural use of pesticides to the environmental impact on aquatic life in regional surface water systems. *Water Sci. Technol.* 49: 125-134.

Karickhoff SW (1981). Semi-empirical estimation of sorption of hydrophobic pollutants on natural sediments and soils. *Chemosphere* 10: 833-846.

Kristoffersen P, Rask AM, Grundy AC, Franzen I, Kempenaar C, Raisio J, Schroeder H, Spijker J, Verschwele A, Zarina L (2008). A review of pesticide policies and regulations for urban amenity areas in seven European countries. *Weed Res.* 48: 201-214.

Mackay D (1991). *Multimedia environmental models: The fugacity approach.* Lewis Publishers, CRC Press, Boca Raton, FL, USA.

Marer PJ (1999). *The safe and effective use of pesticides.* University of California publication 3324, Oakland, USA.

Puijker L, Van Beek K, Beerendonk E, Gijsbertsen A (2004). *Door drinkwaterbedrijven gemaakte kosten als gevolg van bestrijdingsmiddelengebruik. Inventarisatie over de periode 2001–2003* (Costs made by drinking water companies owing to pesticide use). VEWIN and KIWA, KWR 04.094 (in Dutch).

Ramwell CT, Heather AIJ, Shepherd AJ (2002). Herbicide loss following application to a roadside. *Pest Manag. Sci.* 58: 695-701.

Silva E, Batista S, Viana P, Antunes P, Serôdio L, Cardoso AT, Cerejeira MJ (2006). Pesticides and nitrates in groundwater from oriziculture areas of the 'Baixo Sado' (Portugal). *Int. J. Environ. An. Ch.* 86: 955-972.

Spikkerud E (2002). Pesticide Aquatic Risk Indicators: Testing the OECD indicators REXTOX, ADSCOR and SYSCOR and the Norwegian aquatic risk indicator with estimates of use data from Norway. Technical Report of the OECD (Organisation for Economic Co-operation and Development) Project on Pesticide Aquatic Risk Indicators. Available via: *http://www.oecd.org/dataoecd/6/17/2752913.pdf.*

Spliid NH, Carter A, Helweg A (2004). Non-agricultural use of pesticides – environmental issues and alternatives. *Pest Manag. Sci.* 60: 523.

Tomlin CDS (2006). *The Pesticide Manual, 14th ed.* British Crop Protection Council, Farnham, Surrey, UK.

TOPPS (2008). TOPPS Training framework Preventing water contamination from point sources when using Plant Protection Products on arable, fruit and grassland farms. Train the Operators to Prevent Pollution from Point Sources (TOPPs) Tutor handbook. Available via: *http://www.topps-life.org/upl/19/default/doc/080411TOPPS%20Training%20framework%20handbook.pdf.*

Van De Steene JC, Stove CP, Lambert WE (2010). A field study on 8 pharmaceuticals and 1 pesticide in Belgium: Removal rates in waste water treatment plants and occurrence in surface water. *Sci. Total Environ.* 408: 3448-3453.

Wittmer IK, Scheidegger R, Bader HP, Singer H, Stamm C (2011). Loss rates of urban biocides can exceed those of agricultural pesticides. *Sci. Total Environ.* 409: 920-932.

INDEX

B

C

D

J

K

L

M

N

O

P

Q

R

S

V

W

Y

Z